50 Au[illegible]
Auto Projects for the Evil Genius

Evil Genius Series

Bionics for the Evil Genius: 25 Build-it-Yourself Projects

Electronic Circuits for the Evil Genius: 57 Lessons with Projects

Electronic Gadgets for the Evil Genius: 28 Build-it-Yourself Projects

Electronic Sensors for the Evil Genius: 54 Electrifying Projects

50 Awesome Auto Projects for the Evil Genius

Mechatronics for the Evil Genius: 25 Build-it-Yourself Projects

MORE Electronic Gadgets for the Evil Genius: 40 NEW Build-it-Yourself Projects

123 PIC® Microcontroller Experiments for the Evil Genius

123 Robotics Experiments for the Evil Genius

50 Awesome Auto Projects for the Evil Genius

GAVIN D. J. HARPER

McGraw-Hill
New York Chicago San Francisco Lisbon
London Madrid Mexico City Milan New Delhi
San Juan Seoul Singapore Sydney Toronto

The McGraw-Hill Companies

Cataloging-in-Publication Data is on file with the Library of Congress

2 3 4 5 6 7 8 9 0 QPD/QPD 0 1 0 9 8 7 6

ISBN 0-07-145823-9

The sponsoring editor for this book was Judy Bass and the production supervisor was Pamela A. Pelton. It was set in Times Ten by MacAllister Publishing Services, LLC. The art director for the cover was Anthony Landi.

Printed and bound by Quebecor/Dubuque.

This book is printed on acid-free paper.

To the late Ray Cooke, a wonderful man, thank you for everything you taught me about electronics. I owe my passion for the subject (and most of my knowledge) to you.

Ray ran the race with patience, "Looking unto Jesus" (see Hebrews, chapter 12, verses 1 and 2).

About the Author

Gavin D. J. Harper is an automobile hobbyist, member of a number of classic car clubs, and devoted British car enthusiast. His current "fleet" includes a 1983 Austin Mini and a 1979 MG Midget, much to his family's displeasure. Gavin is currently studying with the Open University in the Honors program toward a bachelor of science in technology and with the University of East London toward a master of science in architecture with a focus on advanced environmental and energy studies. He holds qualifications in IT, engineering, and computer-aided design. Gavin is a Science and Engineering Ambassador for the SETNET group and is actively involved in promoting science, engineering, and technology in schools.

Contents

Foreword

I have been lucky to have worked on several prestigious car projects, ranging from Ford Transit to Aston Martin DB7. Although my general interest was on the materials side, I can't help but appreciate the advance that electronics have made. Welcome to the "sophisticated car electronics."

Take one subject, engines. The days of being in a position to correct a problem under the bonnet are now a thing of the past. Problems are now detected via advanced electronic engine management systems, and they are analyzed and decoded to give the reason for the problem. Easy!

The subject of electronics governs nearly all cars. Transmission systems on several cars today have a "fly-by-wire" system that governs gear change, similar to that found on Formula One (FI) cars today. This is no accident. F1 cars are seen by the public as being on the cutting edge of vehicle technology. Car companies that operate today are multinational and cover the globe and a range of car types. Audi/VW, for instance, is one company that people see only one part of. The general public may not know that the same company also controls Lamborghini, Bentley, SEAT, and Skoda. And for the last 5 years, Audi has run and won (nearly) every LeMans race. Car buyers want to see these evolving performance technologies transferred to their four-seater saloon, and they are willing to pay a premium for that privilege.

Advanced electronic sensors are used to give the driver information about the radio station, the CD track, the bass/treble control, and the parts of the car where music may be required and at what volume. Tire pressures are monitored to make sure that the information is available to the driver. Interior temperature control is monitored for the comfort of all passengers. The list of the roles electronics plays in vehicle manufacturing today is endless.

Gavin Harper is the perfect guide and offers the reader an excellent overview on the topic of sophisticated electronics. I first met Gavin about a year ago and was impressed by his level of knowledge and motivation. Gavin has since been introduced to several high-tech specialist motor manufacturers who have requested his help. I cannot praise him highly enough.

I wish him all the best with his endeavours and feel privileged to have known him.

Stephen Scarborough

Acknowledgments

A big thank you goes out to all of my family, for being there throughout the project. Many thanks to my mum for helping me with the flood of administration and filing duties, my father for helping with printing, and my grandparents for tolerating me rumbling around in their attic 'til the early hours of the morning. Plus thanks to all the family members who have encouraged me to be me and pursue my interest in the early days.

There are a number of people who are expressly to blame for nurturing my interest in things that go "broom broom."

J. Williams x3 (that is Jan, John Jr., and John Sr.): Thanks for putting up with the madcap ideas. John Ambrose, thanks for chewing things over at work, saving me magazine articles, and providing inspiration for new ideas. Thanks, John Bullas, owner of the minilist and general fountain of knowledge, for providing technical suggestions and general madness. Thanks also to Alan Massey, who has helped me to understand the "greasy bits" and instilled in me the notion to walk before running.

My thanks go to Werner du Plessis, Colin Brix, and all the other guys at VIA in Taiwan who have been so helpful and to their associates at AutoCOM, especially Hae Mi Pak. Thanks also to Richard Woo at AutoCAN for his product knowledge and help.

My sincerest gratitude to Ronald Hudnell at Rostra Precision Controls, who provided me with a mine of information about his fantastic products. I can highly recommend this company as having fantastic customer service.

A special mention to Armen and Marina at Digital WW, one of the best providers of Car PC parts on the Web. Many thanks for information about their products. These guys cater to some of the niche products that would otherwise be hard to find. The section on installing a touchscreen is based largely on Digital WW's tutorial.

Gratitude to Christopher Obi Okonjo at British Standards Institute for permission to feature BSI information in the Appendix A.

I am grateful to Jose Maria Vazquez at Deluo for allowing me to pick his brains on GPS tracking and guidance.

Thank you to Mike Fahrion and Matt Williams at B&B Electronics who were incredibly helpful with data about the OBD-II standard and Autotap. Without them you probably wouldn't be reading the chapter on OBD II.

And to Ron "Ronabillies" Boyton who encouraged me in IT and was always at hand with a solution where I had created a problem.

I would like to thank the editors, authors, and forum members of MP3Car.com for their contributions to this book. The MP3Car.com team includes Nathaniel H. Wilson, Robert D. Wray, Jason D. Lewis, and Michael A. Hall.

Finally comes the biggest thank you of all: Judy Bass must be the most fantastic person in New York. Without her, this book would not have been published. Since September 2004, she has made this happen! She has kept me on track and enthusiastic, made everything run like a well-oiled machine, and taken care of all the unseen bits that go unpraised. Thanks, Judy, for being Judy! Writing your first book is pretty hard—you don't know what to expect and how long things will take. I consider myself a very lucky person (some would say blessed) that I have such a great editor who has nursed me through the whole process.

Gavin D. J. Harper

Introduction

Rapid advances in electronics have led to the modern automobile becoming increasingly sophisticated. In the past 50 or so years, with the growth of the microelectronic technology industry, we have seen a massive expansion in the number and complexity of electrical devices used in the automobile.

Although a couple of decades ago, a mobile telephone would fill the trunk of a car and cost the same as a small house, today the same accessory is commonplace and fits in the palm of your hand.

Car entertainment has risen from vacuum-tube-based radios to the dizzying heights of modern DVD navigation systems and multimedia entertainment. This has been made possible by the advent of solid-state devices and the integrated circuit.

Vehicle design is now more exciting than ever before, and the wealth of opportunities created by in-car electronics is being exploited by manufacturers to sell their new models. There is no reason why this technology should not be within the reach of the home mechanic, car-modeler, or kit car builder.

This book aims to demystify some of the sophisticated technologies that are presently employed in luxury automobiles and make them accessible to the home-electronics hobbyist.

With cars increasingly based on networks and buses and with so much integrated into the *electronic control units* (ECUs), it is becoming difficult for people to modify their car's electronics system. This book aims to make some of the sophisticated functions that are commonly achieved through microelectronics affordable and within the reach of Joe Public. The book is not intended to be a comprehensive reference piece, nor is it prescriptive in the application of its circuits. Instead, it is written to be as open as possible in its approach so as to be a starting block for big plans.

In writing it, I have tried to include as many links to useful Web sites as possible. Use the book as a springboard for obtaining more comprehensive information.

This book has been written for everyone who loves to tinker with cars, from the kit car builder to the weekend mechanic.

I hope you enjoy it and gain as much fun from constructing the projects as I have.

Chapter One

Instrumentation

Clear, accurate instrumentation is essential if the driver of the car is to have accurate information about the condition of the vehicle and its performance. In the past, mechanical gauges have been employed to give readings of a vehicle's condition. Electrical gauges that require less bulky connections between the "sender" unit and the gauge have gradually superceded these analogue gauges. In a more modern vehicle, digital gauges offer a clear, striking alternative to conventional analogue dials.

Certain gauges are mandatory and must be calibrated within close tolerances to comply with local legal regulations. Others are optional but nice to include. In this chapter we will look at how to build digital gauges that can interpret signals from a variety of sources. After looking at a number of different gauge designs and arming yourself with some fundamental information, you should be able to design your own digital gauges with confidence.

Project 1: Constructing a Digital Gauge

You Will Need

You Will Need

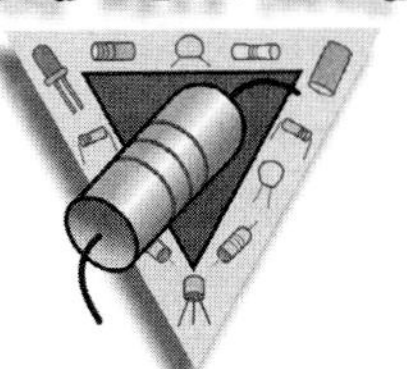

- Clear sheet acrylic, Lexan, or Perspex
- Epoxy resin adhesive
- Fixings to suit
- 12-volt LEDs or small bulbs for backlight
- M3 nuts and bolts

Tools

- Fine-toothed saw
- Vice
- Permanent marker
- Hand drill
- 3mm drill bit
- Countersink drill bit

The projects in this section (with only a few exceptions) culminate in the construction of a digital dashboard gauge.

Although circuit diagrams are fine and dandy on their own, they say nothing of the practicalities of actually mounting the *light-emitting diodes* (LEDs) in the dashboard.

In this project we aim to explore how you might construct a digital gauge. Of course, many possible variations exist for mounting LEDs. This is just one example. You can use the design in its entirety or modify it for your own applications.

For this method of gauge construction, you will need to cut out identical shapes of plastic. The gauges in this book are based on a standard 80mm dimension template. This can be seen in Figure 1-1. You will need three identical pieces of plastic cut from this template. The easiest way to do this is to lay the clear plastic

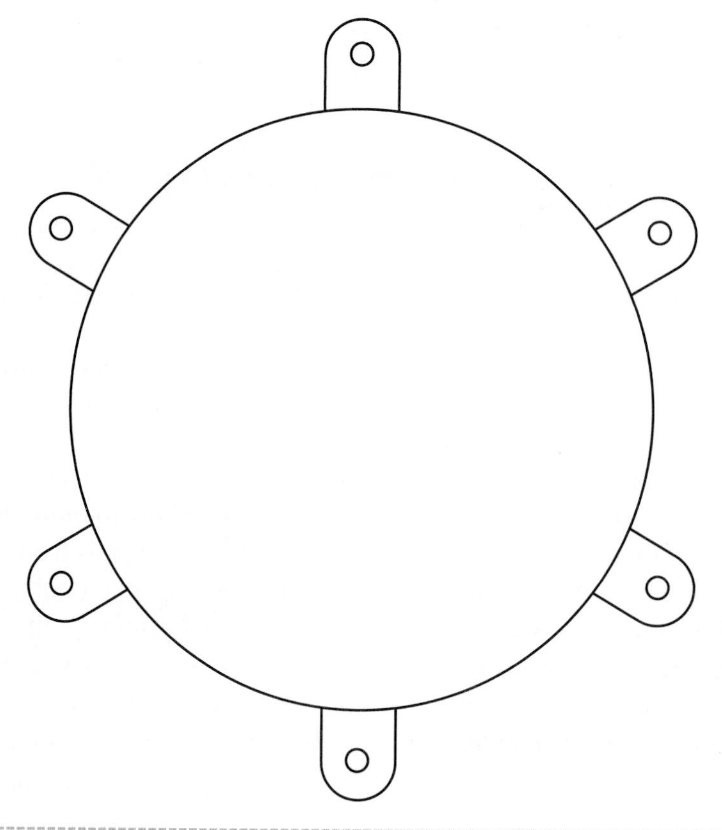

Figure 1-1 *Template for 80mm digital gauge (Courtesy Gavin Harper)*

over the template and draw on the protective film of the clear plastic with a permanent marker. The shapes can then be cut out using a fine-toothed saw.

Do not remove the protective film from the clear plastic until you have finished all cutting and drilling operations.

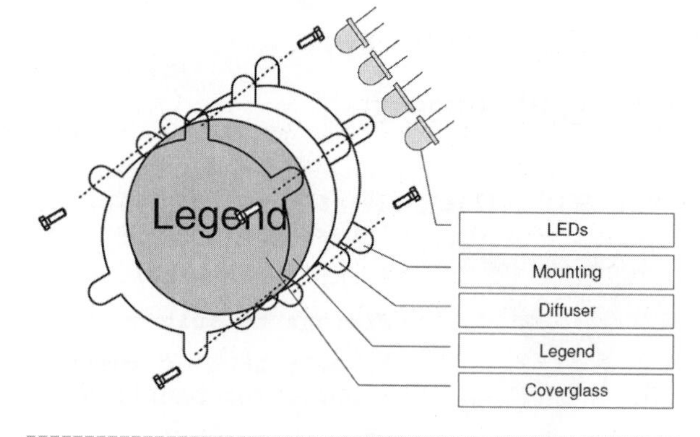

Figure 1-2 *Construction diagram of digital gauge (Courtesy Gavin Harper)*

The gauge is built up in layers. The construction can be seen in Figure 1.2.

Looking at the gauge from the perspective of the driver, the first thing you will see is the protective glass covering the gauge. Six tabs protrude from the large circle. Three of these are used to secure the gauge to the dashboard from behind using screws, nuts and bolts, or similar fixings, depending on the application. The other three tabs are used to assemble the gauge and hold the layers together. The most pragmatic way of doing this is to countersink three screw holes in the cover plate sheet of plastic and then pass an M3 bolt through all of the gauges, securing them at the back with an M3 nut.

Next is placing the design on the face of the gauge. This is done by lithographically printing onto a plastic film (in a pinch, a good-quality thin card can be used). At the end of this chapter, I have included a number of examples of gauge legends for the different projects featured in this chapter. It is very easy to create your own designs in a graphics or *computer-aided design* (CAD) program. The important thing is to achieve a good contrast between the black and clear sections. To get a good-quality lithographic print, either use a laser film and some *overhead projector* (OHP) transparency sheets or take this book or your design into a good printer and ask them to do the work for you.

The backlighting LEDs will be mounted behind the gauge, so it is important that the black sections of the design block out direct light from the LEDs. This will allow the backlighting to be diffused to illuminate all sections of the gauge.

The sheet behind the film can be one of the following:

- Opaque or smoky plastic
- Clear plastic, lightly keyed with fine emery cloth

This pane is used for backlighting the gauge; the opaque or keyed finish helps to diffuse the light. When the backlighting is switched on, all sections of the gauge will be illuminated with a dim glow. To achieve good contrast, a different color of LED should be used for the backlighting and gauge segments.

The final piece of plastic is the mechanical support for the LEDs. This piece is drilled where the LEDs are required and also where backlighting LEDs are to be inserted. The LEDs are then epoxied in.

The existing equipment in your vehicle will largely influence the choice of colors that you choose for your display output. In this book, we stick primarily to green and red, where red indicates a danger condition and green a normal condition. This was done for a number of reasons:

- Because green and red LEDs are very cheap
- Because "red" for danger is largely understood by the general population at large

As mentioned, there is no problem with using amber or even more exotic blue LEDs, cost permitting.

Unfortunately, plastic will scratch a little easier than glass, although it presents a good option to the hobbyist as it is easy to work with. If you find scratches developing on your gauge, use a slightly abrasive plastic cockpit cleaner available from General Aviation to remove imperfections.

Project 2: Adding a Lamp Failure Monitor

You Will Need

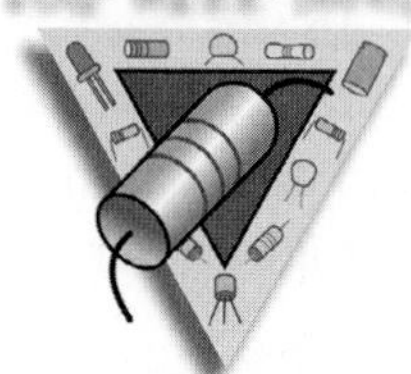

- Reed switch
- Indicator lamp
- Small project box
- Length of wire

Tools

- Soldering iron
- Hot-melt glue gun

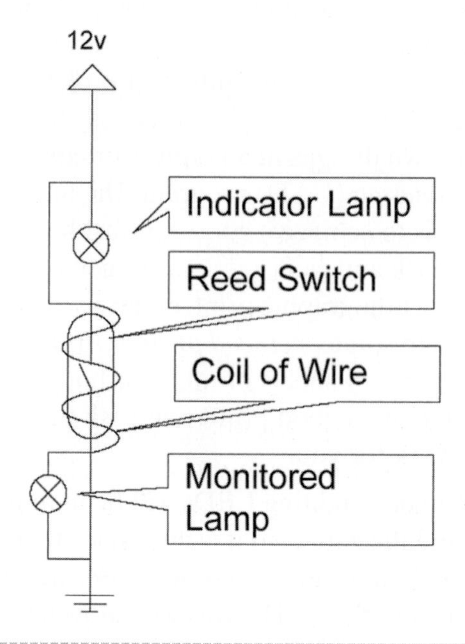

Figure 1-3 *Light monitor schematic (Courtesy Gavin Harper)*

Circuit Operation

When a current flows through the coil, the reed switch closes as a magnetic field is generated. This lights the bulb on the dashboard array, signaling that the lamp is operational. In the event of a bulb failure, the current will not flow and the reed switch will not close; thus, the bulb on the dash will not light. This design is failsafe; if the bulb on the dash fails, it will become apparent when the light to which it is attached is operated.

Construction

The circuit diagram in Figure 1-3 shows the very simple construction of the lamp failure monitor.

The *existing* power supply wire to the lamp is coiled around a reed switch. This assembly should be mounted securely in a project box to provide mechanical support for the reed switch and protect it from damage. If insufficient cable is available, or if the module needs to be removable, cable of a similar or greater cross-sectional area should be used.

The reed switches and cable coils should all be secured in place with hot-melt glue. Be careful not to glue too close to the glass envelope of the reed switch as it could cause the glass to crack with the sudden change of temperature.

You can use a pair of long-nose pliers as a heatsink to dissipate the heat.

CAUTION

Be very careful when handling reed switches—the glass envelopes are *very* fragile!

You will need to construct a suitable mounting for the indicator lamps. Follow the guidelines in Project 1 if you wish to build a gauge cluster for your indicator lamps. One possible variation is to obtain a bird's-eye view of your vehicle, import it into a graphics drawing program, and draw a skeleton of your car. This will form the basis of a legend for a gauge cluster. The appropriate lamp symbols can then be placed in the corresponding positions on the vehicle. Alternatively, single lamps can be attractively mounted on the dashboard in suitable bezels. Figure 1-4 includes some symbols that you may want to use when making a legend.

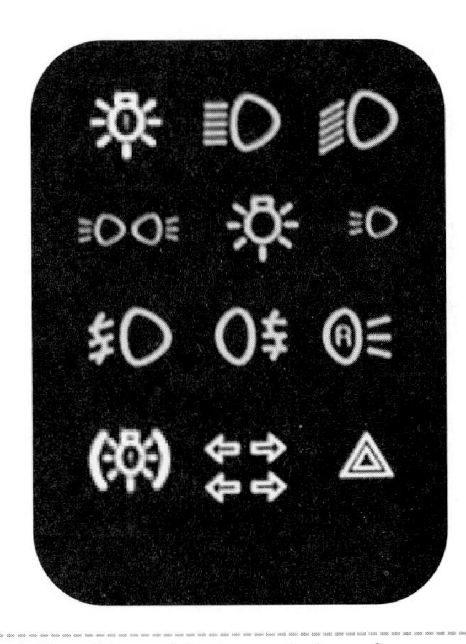

Figure 1-4 *Copyable legend sheet for use with lamp failure monitor circuit (Courtesy Gavin Harper)*

Introducing . . . The LM391x Series of ICs

These ICs are going to become your friends if you want to represent any physical quantities on your car as a bar-graph display.

Implementing them in digital dashboard design really is quite simple. The LM391x series (see Figure 1-5) effectively takes a voltage and displays its value as a number of LEDs. It really is that simple. The beauty of these devices is that they allow the display to show as either

- A continuous bar
- A moving dot display

This gives the designer a great deal of flexibility.

The differences between the ICs are highlighted here. These differences should be considered carefully when selecting an IC for the application:

- The LM3914 produces a linear display in response to the input.
- The LM3915 produces a logarithmic display, in -3dB steps.
- The LM3916 produces a VU scale (-10dB, -7dB, -5dB, -3dB, -1dB, 0dB, 1dB, 2dB, 3dB).

LM391x

Pin	Name	Name	Pin
1	LED1	LED2	18
2	V-	LED3	17
3	V+	LED4	16
4	RLC	LED5	15
5	SIG	LED6	14
6	RHI	LED7	13
7	REF OUT	LED8	12
8	REF ADJ	LED9	11
9	MODE	LED10	10

Figure 1-5 The highly versatile LM391x bar graph display chip pinout

There are a number of common features. For example, the ICs share the same pinout and DIL18 packaging.

So How Do We Go About Building a Digital Dashboard . . .

Digital dashboards in a car get their information from sensors placed around the car. The sensors are connected using some sort of interface to the dashboard display. It is the job of the interface to convert the signal from the sensor into something that can be represented in some way by LEDs or other optoelectronic devices.

In the case of automotive electronics, sensors largely provide a voltage or a resistance. In the case of a resistive sensor, a voltage can be produced easily using a Wheatstone bridge circuit.

Other sensors, such as speed sensors, may provide a frequency output. Quite often a mechanical device is attached to a rotating shaft to help monitor the speed of the vehicle. This mechanical device may take the form of a magnet of some sort of encoding disk. A pickup monitors this mechanical device as the shaft rotates. As the magnet passes a coil or a hall-effect sensor, the magnetic field changes. Similarly, as an encoder disk rotates, light is alternately allowed and blocked. The frequency that the pickup produces is proportional to the speed of the drive.

In the next couple of projects, I am going to talk you through building a variety of gauges. The gauges are all very similar in construction. You will notice that it really is very easy to design your own gauge using these simple building blocks.

For your reference, the manufacturer's datasheets for the LM391x series devices can be obtained from the National Semiconductor Web site at

www.national.com/pf/LM/LM3914.html
www.national.com/pf/LM/LM3915.html
www.national.com/pf/LM/LM3916.html

Using the LM3915/6 Effectively in Your Own Designs

Selecting the Mode

What sort of display you want will determine how you connect pin 9.

- If you want a bar-graph display where LEDs light progressively, then connect pin 9 to pin 3. (It is tied to the positive rail.)
- If you want a dot display, where one LED lights at a time, then leave pin 9 open.

If you want to use the dot display mode with 20 or more LEDs, you will need to connect pin 9 of the *first* LM391x driver chip in the series (i.e., the one with the lowest input voltage comparison points) to pin 1 of the next highest LM391x driver chip. For every other daisy-chained chip, connect pin 9 of the lower input drivers to pin 1 of the higher input drivers for 30, 40, or more LED displays. The last LM391x driver chip in the daisy chain will have pin 9 wired to pin 11. All previous drivers should have a 20K resistor in parallel with LED number 9. This means the resistor should be connected from pin 11 to V_{LED}.

Daisy-Chaining Displays

One of the great things about the LM391x series of chips is that they can be connected head to tail to form bigger, better dashboard displays. In the mixture meter circuit, you can see an expanded scale being used. This is possible with any of the circuits in this chapter. A bigger scale increases display resolution and gives a more accurate output to the driver.

The data sheets give a wide range of information on constructing expanded scale meters.

As well as driving LEDs, the LM391x series can also drive low-current incandescent lamps.

Project 3: Constructing a Simple Voltmeter

This circuit will serve as an ideal and easy introduction to building circuits with the LM391x family of ICs.

We are going to construct a simple voltmeter that will measure the supply voltage of the vehicle and give some indication of battery condition.

You Will Need

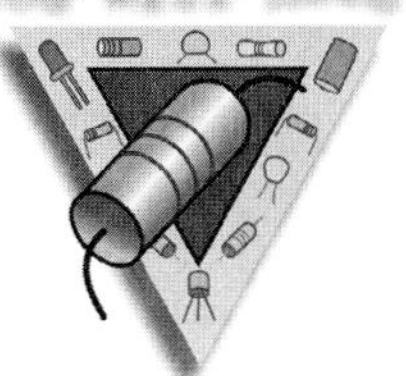

- 1.2K resistor
- 4.7K resistor
- 200R preset potentiometer
- 5K preset potentiometer (×2)
- LEDs, color to suit (×10)
- LM3914 IC

Tools

- Soldering iron
- Side cutters

See Figure 1-6 for the circuit diagram.

Constructing this circuit is a fairly simple affair due to the lack of support circuitry required for the LM3914 IC.

There are two controls for the circuit that you must note. The Offset control moves the scale up and down, and the Span control adjusts the width of values being monitored. This circuit is very simple and is a building block for many simple digital gauges.

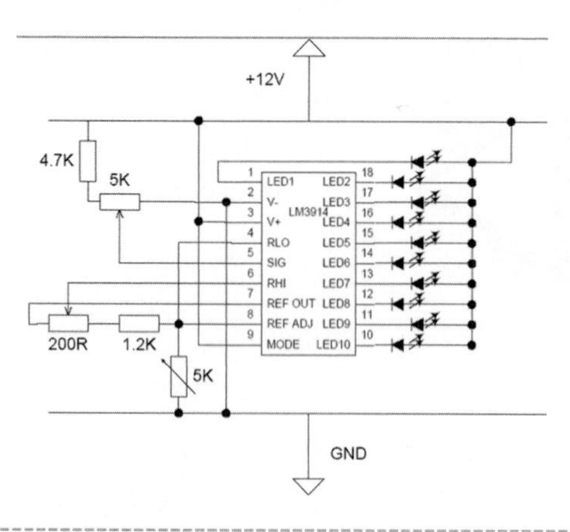

Figure 1-6 *Simple voltmeter*

Project 4: Constructing the Mixture Meter

Many contemporary cars are fitted with an *exhaust gas oxygen* (EGO) sensor. These sensors monitor the amount of oxygen in the exhaust. And from this we can derive how the fuel is burning and hence get a clearer picture of the engine's fuelling requirements.

On cars with a manual choke, a mixture meter can be invaluable for ensuring that the engine is receiving the optimum fuel–air mix. Supplying the engine with what it needs will have positive impacts on both performance and economy.

A mixture meter can be built for less than a fifth of the cost of commercially available units. The circuit built here fulfils the same function as those purchased and can be assembled for very little.

When choosing an EGO sensor, some choices will need to be made. The two main types of sensors are preheated or unheated. For the sensor to work effectively, it must be very hot indeed. This can be achieved either by naturally heating the sensing element with the car's exhaust or by electrically heating the sensing element. Sensor selection is covered in a little more detail in the next project.

The electrically heated sensors are more expensive, but they reach their operating temperature more quickly so you get accurate results in a hurry.

You Will Need

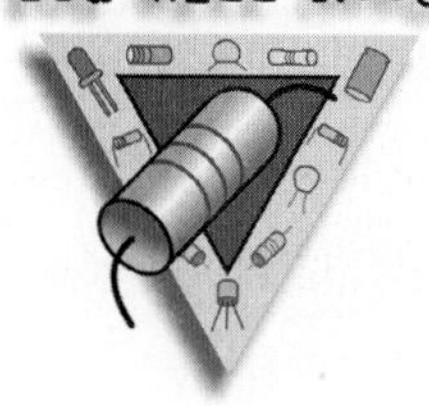

- LM3915 ICs (×2)
- 7805 voltage regulator
- 3mm red LED (×10)
- 3mm green LED (×10)
- 180R resistor (×2)
- 10K resistor
- 47K preset potentiometer

Tools

- Soldering iron
- Side cutters

The circuit diagram for the mixture meter is shown in Figure 1-7.

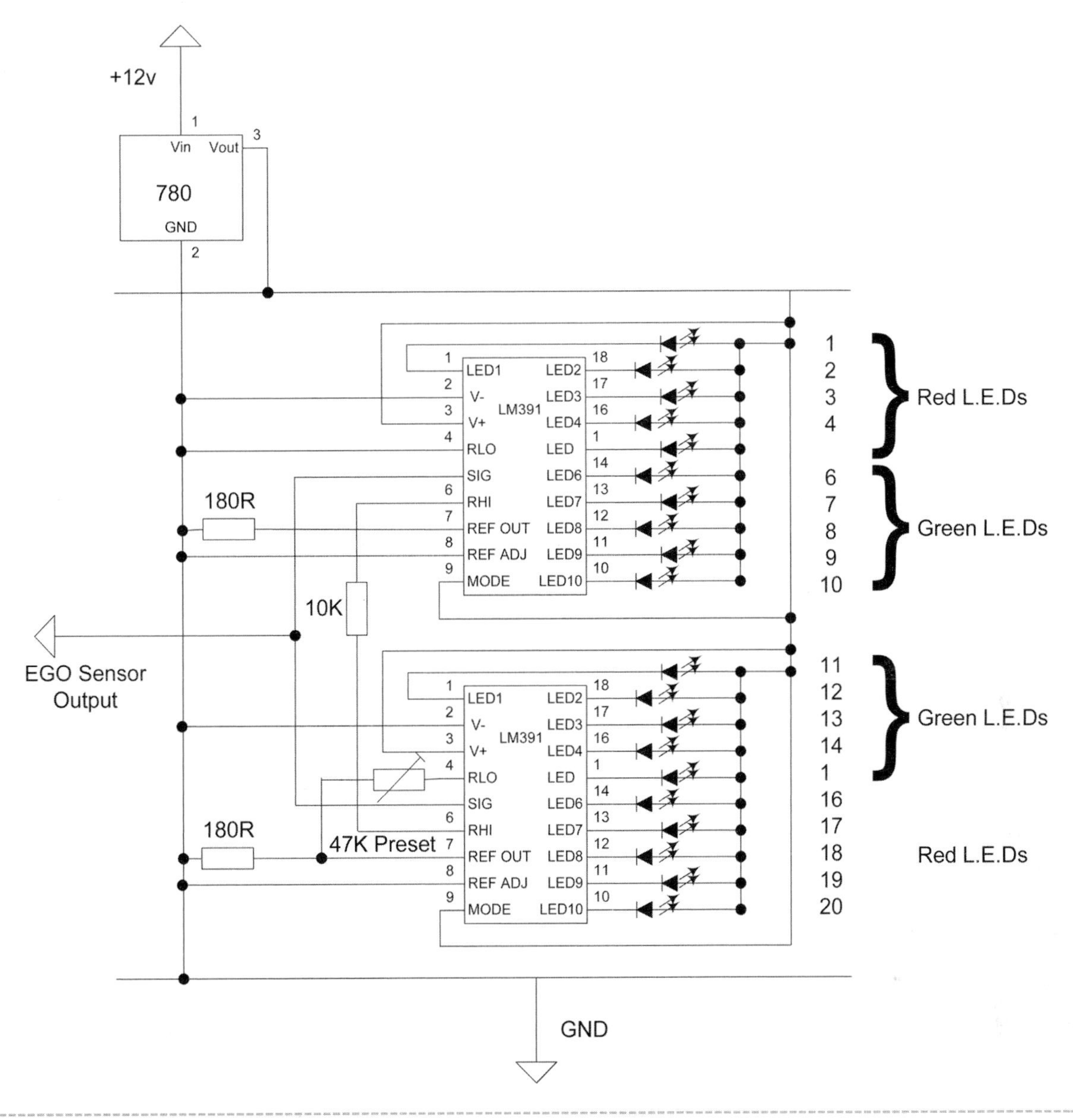

Figure 1-7 *EGO mixture meter circuit*

Construct the ICs on a stripboard or similar circuit fabrication method and use short lengths of wire to interconnect with the display module.

This circuit needs a positive supply, a connection to ground, and a connection to an EGO sensor.

If your vehicle already has an EGO sensor, connect its output to this circuit. If not, follow the next project, "Installing an EGO Sensor."

To ensure when connecting the sensor to the gauge that the wire does not become snagged on high-temperature exhaust parts, route the wire sensibly or use a high-temperature cable.

Once all is connected, move on to the calibration procedure.

Calibration Procedure

Obtain a signal of 0.92 volts. This can be accomplished quite simply with a bench power supply or a variable resistor and 1.5-volt cell.

The voltage needs to be accurate. Check it using a digital voltmeter.

The preset resistor must be adjusted until LED 16 (the maximum-power LED) is illuminated. It will be green in color and at the top of the display.

Troubleshooting

If the project is not working, try connecting a dummy sensor, as shown in the calibration procedure. Adjust the voltage to between 0 and 1 volt. The dot on the mixture meter should move accordingly.

If this works, the problem may be with the EGO sensor. If, however, the lights do not change, check all connections, as well as all soldered joints, thoroughly.

Project 5: Installing an EGO Sensor

You Will Need

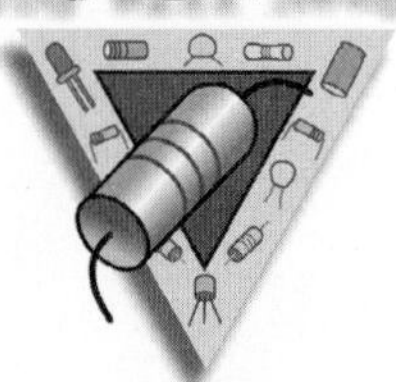

EGO sensor

Tools

M18 tap

EGO sensor socket

The thread for EGO sensors is almost universally M18 × 1.5mm.

Locate the exhaust manifold and pick a point where the sensor will get an average reading of the gases leaving all cylinders.

Ideally, the sensor should be placed after (behind) the manifold where all the pipes converge to a single pipe. The sensor should be mounted before the silencer.

If your vehicle is not fitted with a sensor, it is a matter of drilling a hole in your exhaust after the manifold and cutting an M18 × 1.5mm thread. If you are not comfortable with this, a small, local engineering shop should be able to carry this out. Alternatively, some suppliers of automotive sensors will be able to supply you with a weld-in boss to fulfill this function.

The EGO sensor is fitted to detect the mixture of the fuel that is entering the engine. It calculates this by measuring the amount of oxygen in the exhaust gas. From this information, it is possible to work out how well combustion is occurring and, as a result, derive the fuel–air mixture.

The signal generated is usually fed straight into the car's *electronic control unit* (ECU). It is possible to display the signal, giving the driver an indication of whether the mixture is rich or lean. This feature is especially useful in cars with a manual choke, so the driver can make adjustments accordingly.

A number of different types of EGO sensors are available.

One-Wire EGO Sensor

This type of EGO sensor has one wire, the signal wire, emanating from it. This is the wire that is connected to our circuit. The grounding of the sensor is through the exhaust and the vehicle body. This type of sensor is heated by the exhaust gas.

Two-Wire EGO Sensor (ISO-EGO Sensor)

This type of EGO sensor, sometimes referred to as an ISO-EGO sensor (or isolated EGO sensor) has two wires. One provides the signal and the other is connected to ground. This type of sensor is heated by the exhaust gas.

Three-Wire EGO Sensor (HEGO Sensor)

This type of EGO sensor is self-heating; hence, it is sometimes referred to as a *heated exhaust gas oxygen* (HEGO) sensor. One wire provides a signal to the circuit, while the other two wires are used to heat the sensor to its working temperature more quickly than the exhaust gases would. Grounding for the signal is achieved through the vehicle exhaust and body as with the one-wire sensor.

Four-Wire EGO Sensor (ISO HEGO Sensor)

This type of sensor gives the best of all worlds. It has a heated element and an isolated ground. The four wires comprise a signal wire, a signal ground, and two dedicated heater wires.

Troubleshooting

Out-of-Car Testing and Inspection

If your sensor is not functioning properly, a number of tests can be performed to determine correct operation or the reason for malfunction. The first and simplest test is to remove the sensor from the exhaust and inspect it visually for signs of wear and abuse.

If a shiny metallic deposit is seen on the sensor, check to see if leaded fuel has been used in the vehicle. Leaded fuel attacks the catalyst and renders it inoperative. If this is the case, the sensor should be replaced and only unleaded fuel should be used with the vehicle.

White or gray deposits on the sensor may indicate that oil is being burned by the engine or that additives are being used in the fuel. Certain additives can contaminate the sensor.

Black deposits indicate a buildup of carbon, which can be caused by a rich mixture or by excessive oil consumption due to worn piston rings. Carbon impairs the sensor's operation.

Check also for frayed or damaged wires and dents and knocks to the sensor. Any displaced grommets near the sensor may indicate that wires have become disconnected internally. Usually there are no user-serviceable parts inside the sensor.

In-Car Testing and Inspection

With the EGO sensor still fitted to the vehicle, an oscilloscope can be used to test the sensor's operation.

First, warm the car up to ensure that the sensor is working within the band of its optimum operating temperature.

Then connect the positive lead from the oscilloscope to the sensor output and connect the oscilloscope negative lead to the ground wire or a suitable earthing point as appropriate. A correctly functioning sensor will produce a rapidly oscillating output between 0.1 and 1 volt with the engine running at 2,000 rpm.

Project 6: Building an Expandable VU Meter

You Will Need (Per Channel)

You Will Need

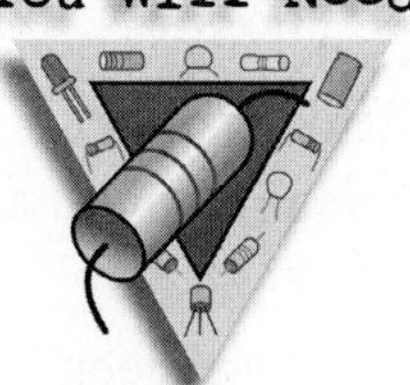

```
LM3915 IC
1N4003 diode
2.2 µF 25-volt electrolytic capacitor
1K resistor
LEDs of any color to suit you (×10)
```

Tools

```
Soldering iron
Side cutters
```

This nifty little VU meter project allows you to construct a "dancing bar graph" for the sound levels in your car. The audio input will accept a standard line-level signal of 1 volt peak to peak with a maximum signal of 1.3 volts. As mentioned in the earlier section, the manner in which pin 9 is configured will determine whether the display is a bar graph or moving dot display. Using the rules in the LM391x section, you can also cascade the chips to provide a greater resolution for your bar graph display.

You could use this project with a conventional car stereo or with a car-PC setup as described later in the book.

A legend is included for a 7.1 surround system. According to your audio setup, you can black out parts of this gauge to form your own custom display. Remember that for a 7.1 ten-steps-per-channel display, you will require eight lots of the components stated in the previous parts list.

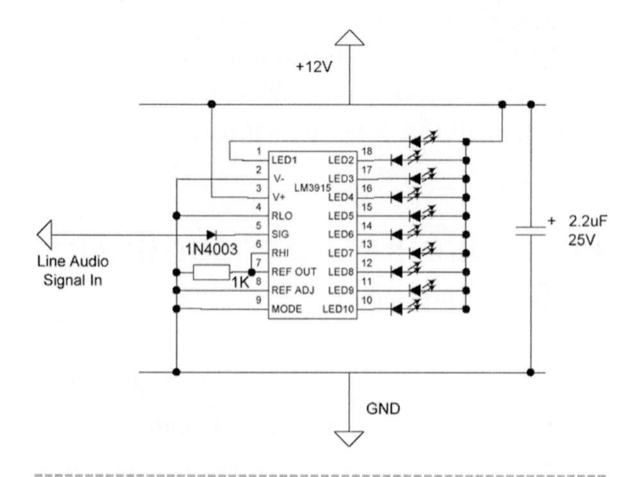

Figure 1-8 *Single-channel VU meter circuit diagram*

Project 7: Building a Digital Tachometer

You Will Need

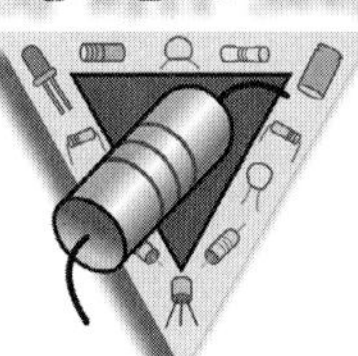

- LM3914 IC
- LM2917 IC
- Zener diode, 12 volts
- 1N4148 diode
- 2.2 µF capacitor
- 22 µF capacitor
- 47 µF capacitor
- 47K (multiturn) preset potentiometer
- 470R resistor
- 1K2 resistor
- 4K7 resistor (×4)
- 5K resistor (×2)
- 10K resistor (×2)
- 20K resistor

Tools

- Soldering iron
- Side cutters

This circuit is a simple tachometer that provides a display that changes in relation to the engine RPM. Earlier in the chapter, we discussed gauges that will accept an input in the form of a frequency; this gauge is one of those circuits. The circuit measures the frequency of the variations in voltage at the negative terminal of the ignition coil.

In a four-cylinder, four-stroke engine, we know that there are two sparks per revolution. As we are working in seconds, and there are 60 in a minute, and as there are two sparks per revolution, we need to multiply the frequency produced by 30 to find the RPM.

The combination of coil and condenser produces quite high voltage fluctuations. Therefore, the first section of the circuit must filter this signal so as not to damage the sensitive ICs.

The first chip in the sequence is the LM2917. This is a frequency-to-voltage converter. It takes the frequency of the ignition sparks and converts it to a voltage that the bar-graph driver can handle. The bar-graph driver then displays this voltage as a series of LEDs. To increase the resolution of the display, simply cascade additional bar-graph drivers.

Project 8: Adding a Voice Annunciator

There is growing concern regarding the distractions and resulting safety issues caused by the increasing number of controls in the modern car. Fortunately, many functions that would traditionally require a gauge or small warning light can be replaced with an audible announcement. This removes another visual distraction from the driver and allows him or her to make better use of all his or her senses when driving the vehicle. The driver's vision can remain focused on the road.

This circuit utilizes the versatile range of an ISD single-chip recording solution. It avoids the complexity associated with a PC- or microcontroller-based solution and works well as a stand-alone system.

The preferred message can be recorded on the chip. This can be a natural human voice rather than a harsh synthesized voice. Or it could be a bell or a chime—any sound that you can produce can be recorded on the chip. Every time that the button is pressed or relay contacts closed, the voice prompt is triggered.

In this circuit, I have used a relay to trigger the circuit. The rationale is simple: This circuit can in many cases replace a 12-volt signal lamp directly. However, many more ways exist to trigger the voice prompt, the only requirement being that the playback contacts are closed.

You Will Need

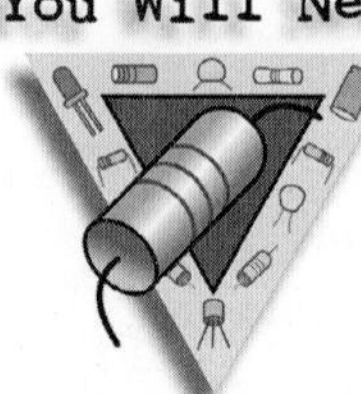

- 1K resistor (×2)
- 5K resistor
- 10K resistor (×2)
- 470K resistor
- 0.001 μF capacitor
- 0.1 μF capacitor (×3)
- 4.7 μF capacitor
- 220 μF electrolytic capacitor
- 16-Ohm speaker
- Electret microphone
- Red LED
- 7805 voltage regulator
- ISD11xx IC
- 12 V relay
- Momentary SPST switch

Tools

- Soldering iron

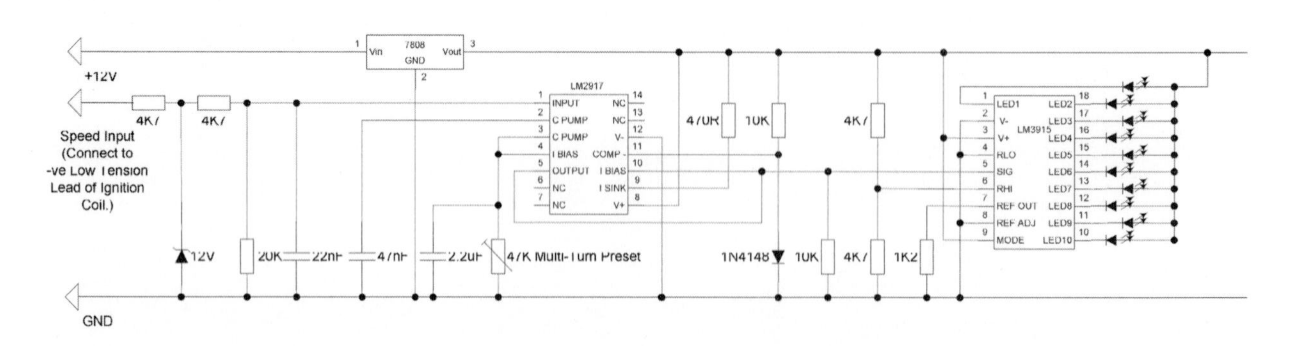

Figure 1-9 *Digital tachometer schematic*

Construction of the circuit can be seen in the circuit diagram (see Figure 1-9). Operation is very simple.

To record a message, depress the record button, and then speak or make a sound into the electret microphone. This sound is digitized and stored in nonvolatile memory. Every time the circuit is triggered, this sound is replayed. Many possible variations exist for this circuit. Another option is to disconnect the microphone and the speaker, and then to input the recording via the *analogue in* (ANA IN) terminal. Similarly the sound output does not have to be via a 16-Ohm speaker; the output can be routed to an amplifier or other audio setup via the *analogue out* (ANA OUT) terminal.

For further information on the ISD 11xx range of chips, the manufacturer's datasheet can be obtained from: www.winbond-usa.com/products/isd_products/chipcorder/datasheets/1100/ISD1100.pdf

Figure 1-10 shows a selection of legends you may wish to use with the gauges in this project.

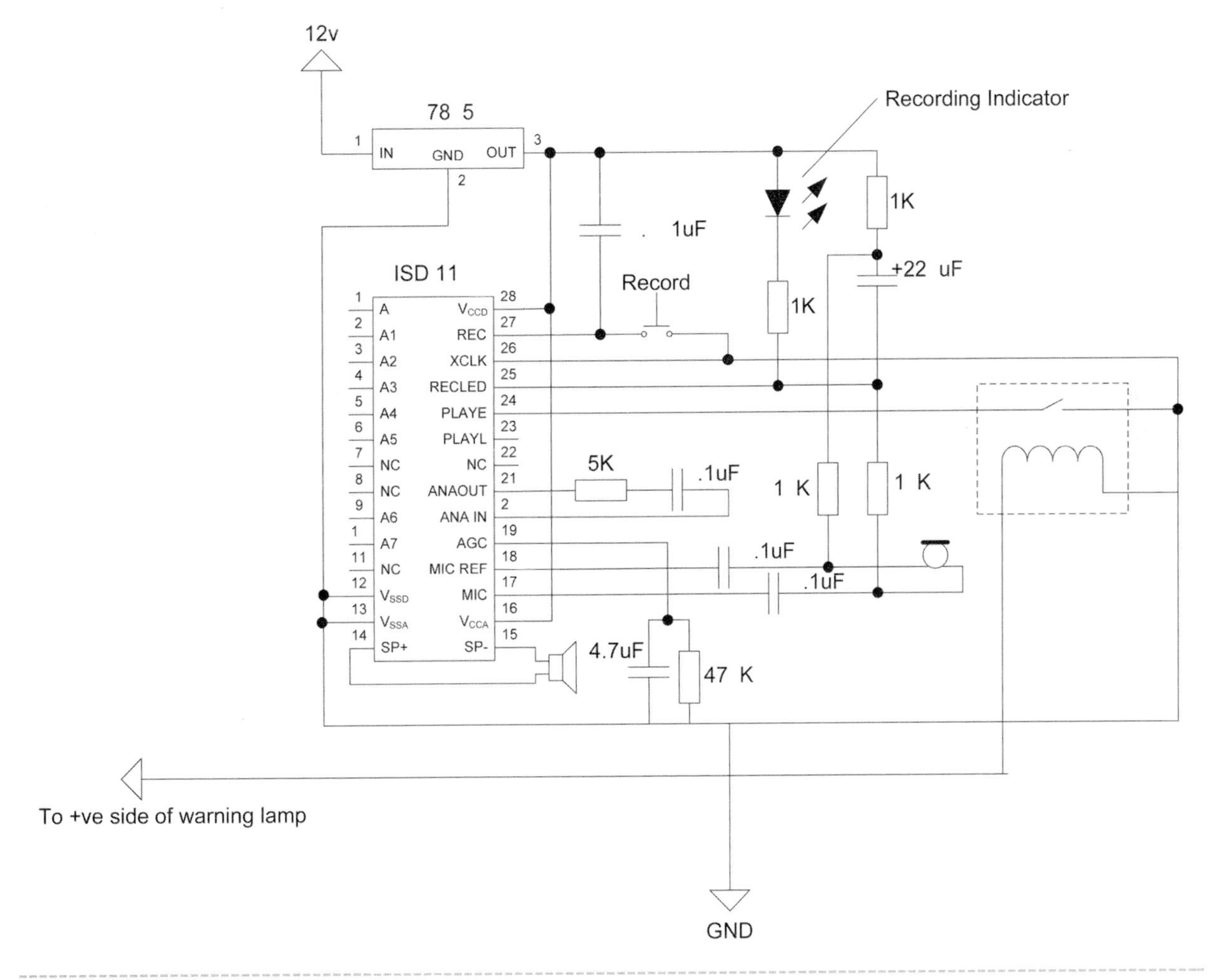

Figure 1-10 *Voice annunciator schematic*

Chapter Two

Comfort Features

Electric Door Mirrors Explained

A feature found on many luxury vehicles is the addition of electrically operated door mirrors. This is easy to achieve with aftermarket products or *original equipment manufacturer* (OEM) mirrors where available.

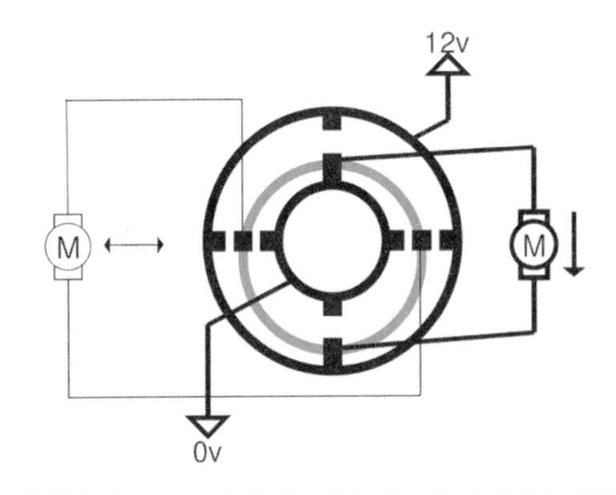

Figure 2-3 *The electric mirror circuit at "down"*

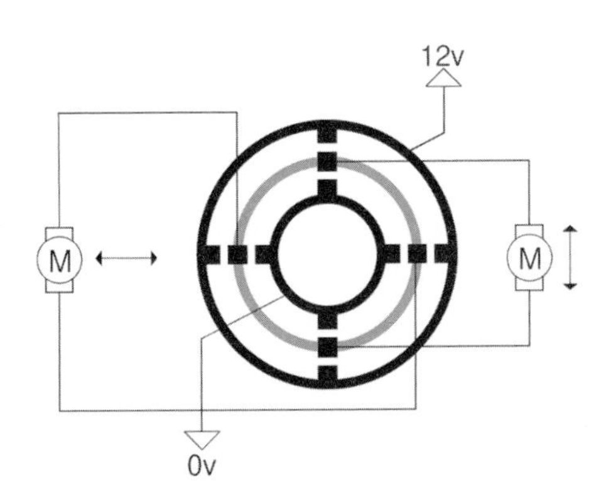

Figure 2-1 *The electric mirror circuit at rest*

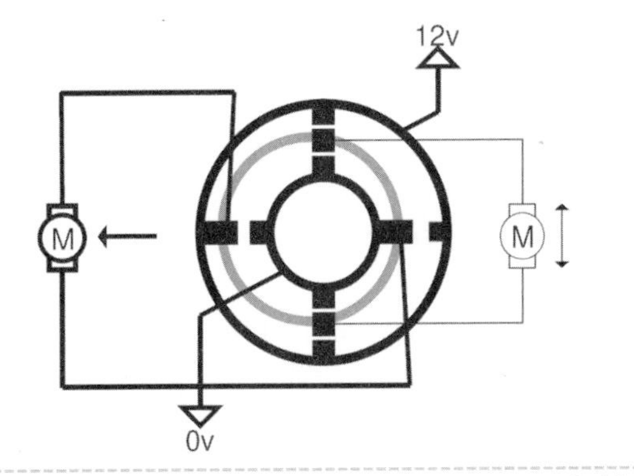

Figure 2-4 *The electric mirror circuit at "left"*

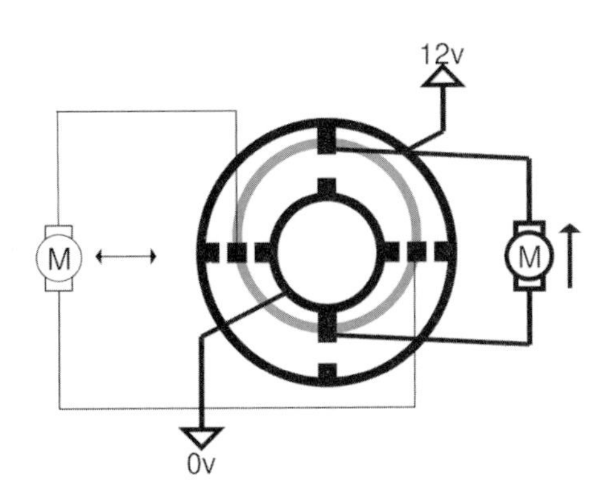

Figure 2-2 *The electric mirror circuit at "up"*

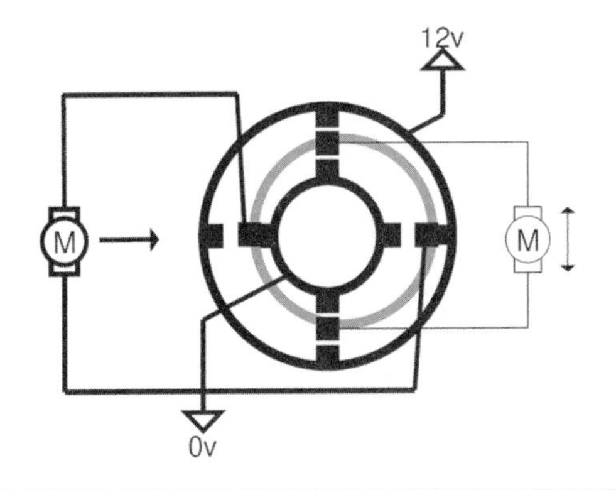

Figure 2-5 *The electric mirror circuit at "right"*

If your vehicle was available with electric mirrors as an option but they were not purchased with the vehicle, it should not be too hard to add them once a fundamental understanding of how the mirror's operate is gained.

The easiest way to control the mirrors is by a joystick control inside the vehicle, mounted at a convenient place for the driver. The joystick operates in the following manner and is illustrated in Figures 2-1 through 2-5. Only one mirror is shown for clarity.

The mirror has two motors inside; one controls the horizontal angular position and the other controls the vertical angular position. Changing the polarity of the supply to the motors changes the direction in which they move.

Inside the joystick are two fixed contacts in the form of rings. One of these rings is connected to a 12-volt potential, the other to the chassis or ground. A cage with four contacts is free to move within these rings. The vertical mirror motor is connected to the top and bottom contacts; conversely, the horizontal is connected to the left and right contacts.

The geometry is such that when the joystick is pushed in a direction, two contacts of the switch cage will contact the rings, one touching the positive and the other touching ground.

Depending on whether the switch has a gate or not, the mirrors may be able to move both horizontally and vertically at the same time. If there is a plastic gate (somewhat akin to your gear selector gate), then you will be able to perform only one movement at a time. It is common practice to use one joystick with a selector switch, which selects the driver's or a passenger's mirror.

Project 9: Installing Electric Door Mirrors

You Will Need

You Will Need

- OEM electric door mirrors or aftermarket door mirrors (with fitting kit)
- A quantity of stranded cable
- Grommets for door penetrations

Tools

- Drill
- Center punch

The mechanical fitting of the mirror to your vehicle will differ largely depending on the vehicle and mirror design.

With many modern aftermarket mirrors, a base plate is available that allows the universal mating end of the mirror to fit to a wide variety of vehicles and to have a custom OEM appearance. The adapter specific to your car is designed to fit in the corner of the window in much the same fashion as the mirror you are removing. The universal mirror will fit to this plate using simple fixings such as nuts and bolts.

With OEM electric door mirrors, the job should be easier by remembering the watchwords "assembly is the reverse of disassembly." When removing the previous mirrors, you may find that you have Bowden cables to remove if the mirrors were mechanically controlled. When ordering the mirrors, try to obtain the modesty plate that fits inside of your vehicle, because when you remove older mechanical joystick mirrors, an unsightly hole will be left. By obtaining the modesty plate, an OEM factory-fresh appearance can be obtained without having to seal off the old hole.

If you are using mirrors with a custom-fit plate for your vehicle, then follow the instructions for fitting that are supplied with the plates. The plates will go in the same place as factory-standard mirrors and will not require further positioning. If you are fitting a universal mirror, follow these instructions.

While you are inside the vehicle and with the car window down, position the mirror on the outside of the door so that it can be clearly seen from the driver's seat. Have an assistant mark the position of the mirror on the door.

Center punch and drill the holes in the door for the mirror. In addition to the fixing holes, drill a hole for the power-supply wires.

A little masking tape on the car body will prevent scratch damage from the drill.

Remove the interior door trim and secure the mirror to the vehicle using appropriate nuts, bolts, and washers.

Now drill a hole in the door and the slam panel to accommodate the mirror wiring.

You will need to choose a position for the control joystick. The a-panel, the door armrest, and the door card could all provide appropriate locations.

Wire the mirror in accordance with Figure 2-1.

CAUTION

Regardless of how the mirror physically attaches to the vehicle, you need to ensure that the cable can run safely through the door. This means the following:

- Use grommets at all penetrations.
- Ensure that the mirror cables will not snag on a window winder or any moving part inside the door.

Now connect the power supply and have fun playing!

For comfort and convenience, many vehicles now come with electric windows. A small, unobtrusive button is much more pleasant and easier to operate than a large handle that must be cranked. In the past this was a luxury feature and was present only in high-end cars, but now it is common to most vehicles.

There are a number of ways to attack the problem of installing an electric window where there was once a manual, or crank, window. Before looking at the differences between the different methods, let's look at the main similarity—the wiring.

The simplest way to power electric windows is directly through a polarity-reversing switch. The switches generally have momentary action, with three `ossible positions. When the switch is pushed one way, the power is connected to the window motor. When the switch is pushed the other way, the power is connected to the window motor, but with the polarity reversed. When the switch is in the center, it is off. At a simple level, this type of control can be achieved with a DPDT (from *dual-pull dual-throw*) switch wired correctly.

The next advance of this method is to use custom electric window switches. These are available at a wide range of auto parts stores and come in a variety of styles, with and without illumination.

The beauty of these switches is that they can be daisy chained, allowing the driver as well as the passengers to control the front passenger and rear windows.

A generic diagram for the wiring of electric windows is shown in Figure 2-6.

In a four-door car, the driver will have a console with four switches, one for each of the four electric windows.

The switches have five connections. An easy way to think of this is two inputs, two outputs, and a ground. The switch is arranged so that when in the resting position, the two inputs are connected directly to the two outputs. When the switch is pressed one way, one of the outputs is connected to the ground while the other output remains connected to its respective input. When the switch is pressed the other way, the reverse is true. The other output is connected to the ground while the remaining output is connected to ground. The first switch in the chain has both inputs connected to a fused positive feed. To have multiple switches controlling one motor, the switches are daisy chained, output to input, with the center contact being the ground. It is possible to achieve a hierarchy of command with the switches by providing a child lock switch to

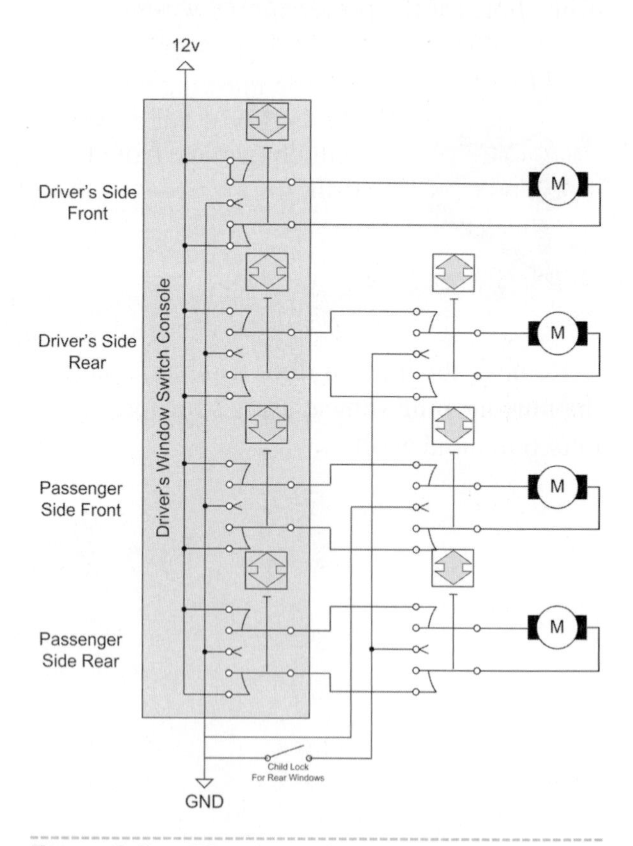

Figure 2-6 *Electric windows wiring schematic (for four-door vehicles)*

disconnect the ground from the rear seat window switches. This prevents tampering with the windows via the switches at the rear windows and avoids accidents.

To enable smaller cables to be used in interconnections between switches and motors, a pair of relays can be employed near the motor. These are supplied by a common feed. This approach reduces wear on switch contacts.

If your vehicle could have been supplied with electric windows as an optional extra, you may be able to source the mechanism from a scrap yard. If you can find one of the mechanisms that was actually designed for your car, you can perform a straight swap.

Because of the broad range of applications, it would not be feasible to discuss the installation of OEM electric windows. We will instead focus on the mechanical installation of aftermarket kits. The electrical installation for OEM windows will be broadly similar if made with point-to-point connections. Note that the layout of an electric system is likely to differ dramatically from a hand-crank window system. You should also be able to obtain the door panels and other interior trim to make the job look factory fresh.

Some OEM window installations will not use point-to-point wiring. In an attempt to reduce overall vehicle wiring costs, a *controller area network* (CAN) system is used whereby switches and motors are all driven by controllers connected to a network. We will not discuss CAN systems in this book.

Three mechanisms are commonly used in electronic windows. The first has an electric motor driving a spur gear, which in turn rotates a large segment of the gear. This is connected to the window lift mechanism. The second mechanism uses stiff, taught cables that hoist the window up and down with a pulley system. The third type of system roughly corresponds to the "Type 2" aftermarket system.

If the originally fitted motors are not available, consider one of the wide variety of aftermarket electric window kits. In many cases, an aftermarket kit will be preferable, as they come with clear instructions and a prescribed method of fitting.

The type of kit can be broken down roughly into three groups:

Type 1 Aftermarket Window Kits

In this type of window kit, a small slim-line gearbox fits over the stub protruding from the window gear where a handle would normally attach (see Figure 2-7). Power is transmitted to this gearbox by a motor, which is mounted remotely at the bottom of the door. The motor drives a flexible cable that turns in a sheath. This sheath is anchored securely at the motor end and the gearbox end with the cable turning within. When buying a new kit, a variety of plastic adapters is supplied to enable you to interface the gearbox with the stub protruding from the window mechanism. The contents of the kit are illustrated in Figure 2-8.

This design has the advantage of being slim, requiring only a very shallow bulge in the door

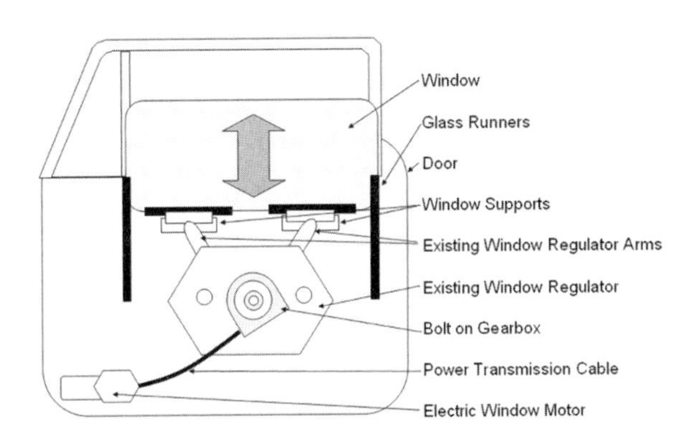

Figure 2-7 *Aftermarket electric windows type I*

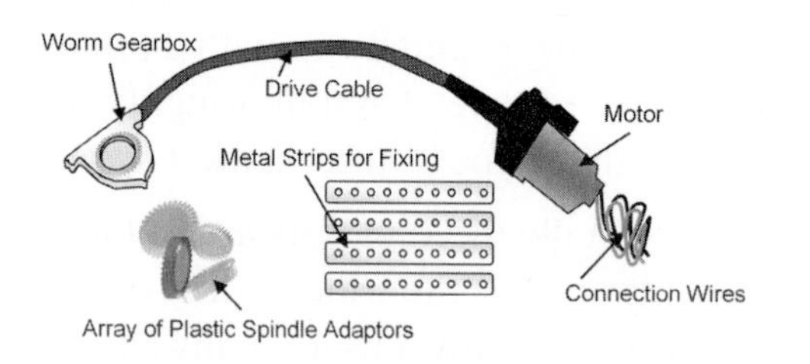

Figure 2-8 *Contents of type 1 aftermarket window kit*

card to accommodate the mechanism. Often kits are supplied with a small binnacle that fits over the hole where the window winder would previously protrude. For custom cars, a fiberglass pod can be made that integrates more evenly with the rest of the car. The disadvantages of this type of mechanism are that the window winding gear is retained, adding weight and mechanical complexity to the car.

Type 2 Aftermarket Window Kits

In this type of electric window (see Figure 2-9), the whole window winder and assembly is removed, leaving only the glass in the door. Where the winder mechanism would previously fit into the door, a riser channel is fitted that attaches to the door with universal fitment brackets. The window is secured to a glide pad that moves up and down the riser. A motor that is mounted remotely provides

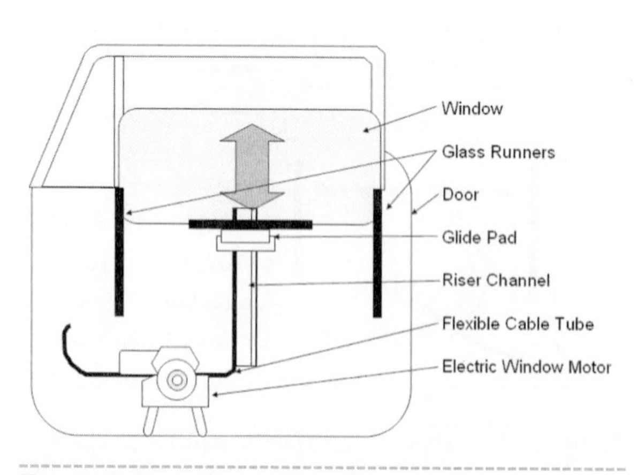

Figure 2-9 *Aftermarket electric windows type 2*

power. A gearbox converts the rotary motion of the motor into a high-torque linear motion, which pushes a flexible rack up and down the riser, providing kinetic power to move the window. The advantages of this installation are that it is light, mechanically simple, and unobtrusive. The disadvantages are that it is one of the most complex kits to fit.

Type 3 Aftermarket Window Kits

This type of kit is very simple to fit (see Figures 2-10 and 2-11). The window winder is removed, and a large pod is fixed to the door over the position where the winder was previously placed. This pod contains the motor and switchgear. The motor directly drives the stub of the window winder. Although this kit is very easy to install, it is far from unobtrusive and still retains the mechanical window winder mechanism, adding complexity to the installation.

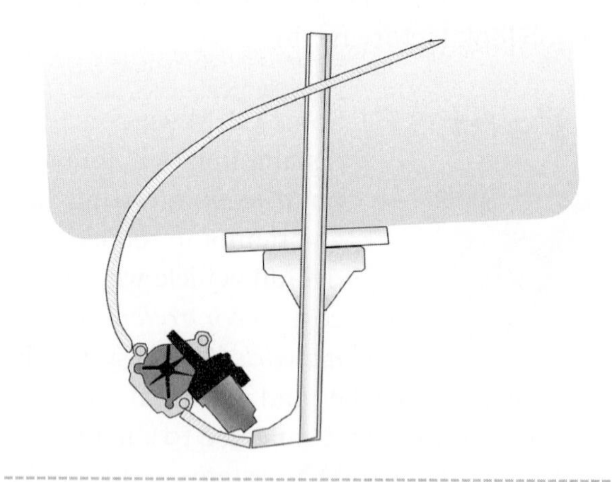

Figure 2-10 *Type II window detail*

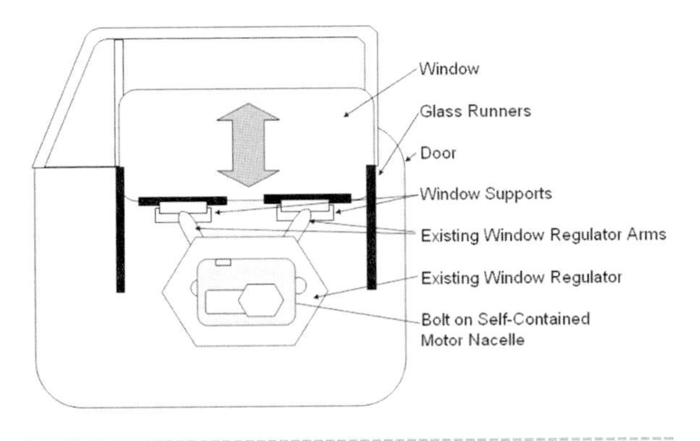

Figure 2-11 *Aftermarket electric windows type 3*

You Will Need

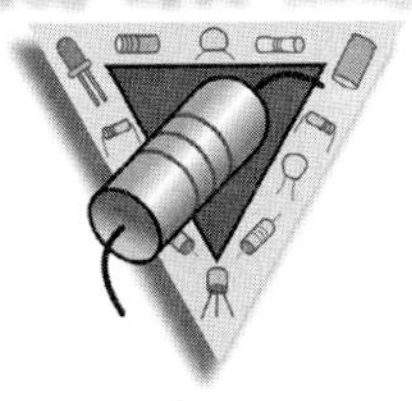

Universal window kit

Tools

Screwdriver selection
Hacksaw
Side cutters
Soldering iron
Drill
Array of drill bits

Mechanical Installation

Step 1

All Types

Remove all door handles, knobs, armrests, switches, and marker lights before removing the door trim.

Step 2

Types 1 and 2

Remove the trim. The trim will be fixed in one of a variety of ways. If clips are used to fix it to the doors, be sure to pry the trim off carefully without breaking the trim clips. If any clips work loose, be careful to retain them for later use. With the doorcard removed, take some time to ensure that the mechanism has enough room to fit.

Type 3

Remove the case from the electric window motor. Move the drive unit up so that it engages with the stub sticking out from the winder mechanism. Mark the position of the mounting holes.

Step 3

Type 1

Look at the spindle adapter protruding from your window winder. Examine the motor worm-drive gearbox. It is quite likely that a number of adapters will be supplied with your electric window kit. Check to see that one fits. It is likely that a small amount of modification will need to be done to the window winder in order to get it all to work.

Type 2

Measure the height of the window opening. Be sure that the kit that you purchase has enough height on the riser channel to allow your window to be raised and lowered completely.

The glide pad will require about 1 inch (25mm) depth from the door to the riser channel. The motor may require 6 inches (150mm) for mounting.

Type 3

Drill the holes for the winder unit, ensuring that no cables or mechanisms behind the door card are damaged in the process.

With the trim panel removed, now might be a good time to check all of the door mechanisms, locks, and so on, to ensure everything is working correctly, and to apply lubricants to parts where required.

Step 4

Type 1

Drop the motor in the bottom of the door where you expect it to be located. It does not matter now if it is not in exactly the correct place — it is highly likely some form of adjustment will be required.

Type 2

Before removing the winder mechanism, tape the glass securely in place with duct tape. This will act as an extra pair of hands while you are working on the mechanism and help to prevent any damage to the glass.

Type 3

Remove the trim panel (as in Step 2 for Type 1 and 2). Drill holes for mounting the unit and for the wiring.

Step 5

Type 1

You will need to secure the gearbox to the window winder. A torque reaction strap will need to be fitted. This is to prevent the gearbox from turning as you wind the window up and down. To get a correct idea of the orientation of the strap, draw a line in pencil from the center of the window winder to the floor. Now draw a line at 90 degrees (perpendicular) to this. You will need to drill a hole in the door panel to secure this strap.

Type 2

Remove the winder mechanism bolts first of all. The mechanism will probably also be secured to the door with some kind of adhesive putty. Make sure all of this is cleaned off. Getting the mechanism out of the window channels requires that it is fully wound to its up position. You can then slide the two arms out of the window channels and remove the mechanism, which can be set aside.

Type 3

Drill holes in the door for the cables.

Step 6

Type 1

Depending on the construction of your vehicle, you may need to drill a hole in the door panel to accommodate the drive cable to the motor from the gearbox.

Type 2

Hold the riser in the place where you anticipate it will go. You will find that for universal fitting kits, the risers are generally too long for your specific vehicle application. Remove all brackets and endstops and then trim the riser to size using a hacksaw.

Type 3

Refit the trim with the motor in place.

Step 7

Type 1

Now comes the time to install the motor! Some motors will have a number of screw holes built into the casing, in which case installation is a relatively easy affair. Failing this, a strap can be bolted around the motor with an insulating strip to ensure the motor does not ground against the door.

Type 2

Look at how the glass connects to the mechanism. It is likely that you will have a roller channel (used by the old mechanism). This is then connected to a sash, which is secured to the window usually using putty or glue.

You will find a new length of roller channel in your window kit. This needs to be cut to the same length as the sash affixed to your window.

Type 3

Replace the cover on the motor. Installation of the mechanical components is now finished; proceed to electrical installation.

Step 8

Type 1

Now the mechanical installation is finished; secure the window gearbox adapter and commence with the electrical installation.

Type 2

Draw a vertical line on the inside of your door using a permanent marker. This will ensure that when you mount the riser channel, it will be plumb so that the window can move easily.

Step 9

Type 2

A roller bar will fill in the channel where the rollers of the window mechanism rested previously. The glide pad can then be fixed to the roller bar. Slide in the riser channel. You will need to fix the riser channel to the door body securely; the kit should supply nuts and bolts for this purpose. You may find that in order to access nuts and bolts you will need to cut or drill holes in the door metal to make them accessible. All holes will be covered by the trim panel when it is replaced.

Step 10

Type 2

Once the window mechanism has been mounted, you need to mount the motor securely to stop it from rattling about. From one end of the motor, there will be a sheath to house the unused rack that pushes the window up and down. This can be secured out of the way with cable ties, but all bends in the sheath should be kept as smooth as possible and kinks avoided at all costs.

Electrical Installation

The electrical installation is the same for all three types of kits. A generic wiring diagram can be found in Figure 2-6.

Important points to remember include ensuring that wherever a cable penetrates a bulkhead it is insulated by a grommet, and the switches will need to be mounted either on the door cards or in a center console.

If you find that the window is traveling excessively fast in the downward direction or that the motor is struggling by trying to push the window too far down into the door, it is a simple matter of fitting an SPDT switch into the downward circuit. When the window touches this switch, the changeover goes from connecting the down motor leg from the down connection of the switch to a positive supply. By doing this, the only way the window can now go is up by grounding the up leg of the motor.

Troubleshooting

When looking at why electric windows fail, consider how well they were performing before the failure occurred.

If the window motors come from a new kit and they have never worked, then it is highly likely that the problem is with the wiring or switchgear.

If the windows are installed and fail progressively over time, moving more slowly and laboring each time they are rolled up, then the problem will quite likely lie with the motor. It may be failing with age or under excess load. The motor may be struggling because the mechanism needs lubricating, or it may be that the particular electric window motor kit is too "light" for the application.

If, however, the failure of an electric window is sudden, then it is more likely to be loose wiring or a defective connection.

When checking connections, first remove the window switch from its housing. Check that power is reaching the switch. Turn your vehicle's lights on and off. Does the voltage level change significantly? (This last step is to ascertain that the power supplied was for the windows and not for illumination of the switch.)

If, when pressing the button, the window does the opposite of what is expected (that is, when pushing down, the window rolls up), reverse the connection of the wires to the motor as the polarity may be reversed.

Project 11: Installing Heated Seats

Another luxury item seldom found previously as standard on vehicles is heated seats. It can be a real boon in the winter to get into a car that feels warm and cozy. Leather and vinyl seats especially can be very cold in the winter and can prove to be uncomfortable until they warm up. And sufferers of back pain may find therapeutic benefits from the gentle, warming, soothing action.

This project focuses on installation of a seat heater kit, as supplied by Rostra Precision Controls (see Appendix A for contact information).

You Will Need

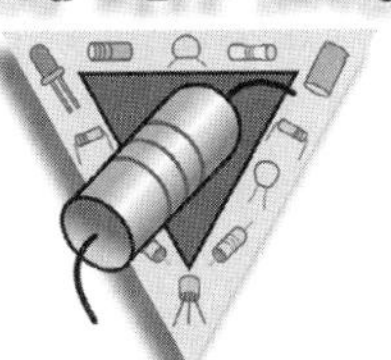

- Seat for modification heater element*
- Controller harness*
- Extension harness*
- Fused harness*
- Switch assembly*
- Control module*
- Bag of miscellaneous parts*

*All of the items with an asterisk are included in the kit from Rostra Precision Controls that you see in Figure 2-12.

Tools

- Drill
- 21mm or 13/16-inch drill bit

Before starting, remove the seat from the vehicle. The process for this will differ from vehicle to vehicle, the details of which could be found in a Haynes™ manual or a similar source.

Seats differ widely in their construction. Before attempting this project, familiarize yourself with how your car's seat is made and reassure yourself that you can competently disassemble and reassemble the upholstery of your seat.

If your seat is of a bonded foam construction, you will be unable to proceed with this project.

The fabric, leather, or vinyl of your seat will be secured using one of a number of methods. Seat upholstery secured using zippers is probably the easiest to remove. Other methods of securing upholstery include the use of J clips, which will spring off when pried; removable-pocket construction, where the seat is held together by pins and screws; rigid back construction, where removing the hard plastic back will allow upholstery to be removed; and finally the uce of "hog rings," which must be removed and replaced using hog ring pliers (an alternative is to substitute the rings with cable ties).

Take great care when removing fixings not to damage the seat fabric.

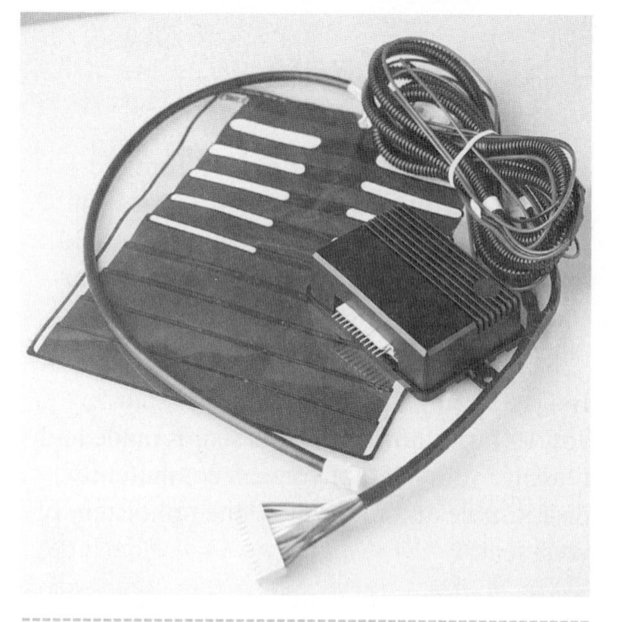

Figure 2-12 *Seat heater kit (Courtesy Rostra Precision Controls)*

Once the seat upholstery has been removed, carefully roll it up inside out to facilitate easier reinstallation.

You will then need to install the heating element. In the Rostra kit, the heating element comes with preaffixed double-sided tape to help secure the element. Both heating elements provided in the kit are exactly the same. One should be used for the lumbar region and the other for the bottom of the seat.

Try to orient the heating elements so that the cables can be routed out through the crease in the rear of the seat. The heating element should be kept away from any metal or plastic seat components.

When positioning the elements, be sure that there is no way it can be folded, cut, or damaged in any way.

At this point, the 10-pin connector on the wiring harness should be plugged into the controller.

You now have an option: The switch that controls the seat heater can work in one of two ways. It has two on positions with a central off. You can see the switch in Figure 2-13.

- In Mode I, the switch is used to select between low and high heat in all regions of the seat.

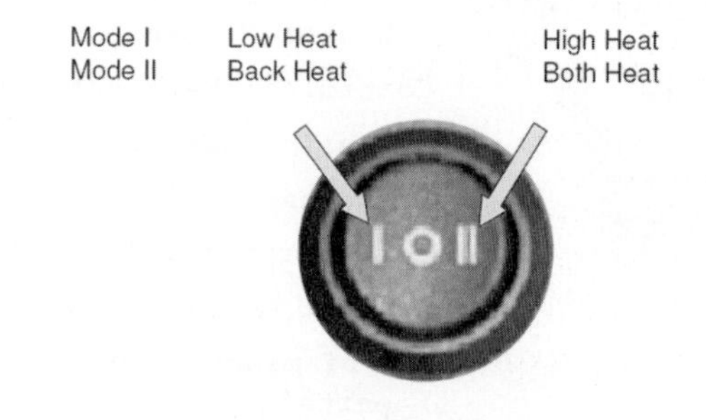

Figure 2-13 *Seat heat selector button (Courtesy Rostra Precision Controls)*

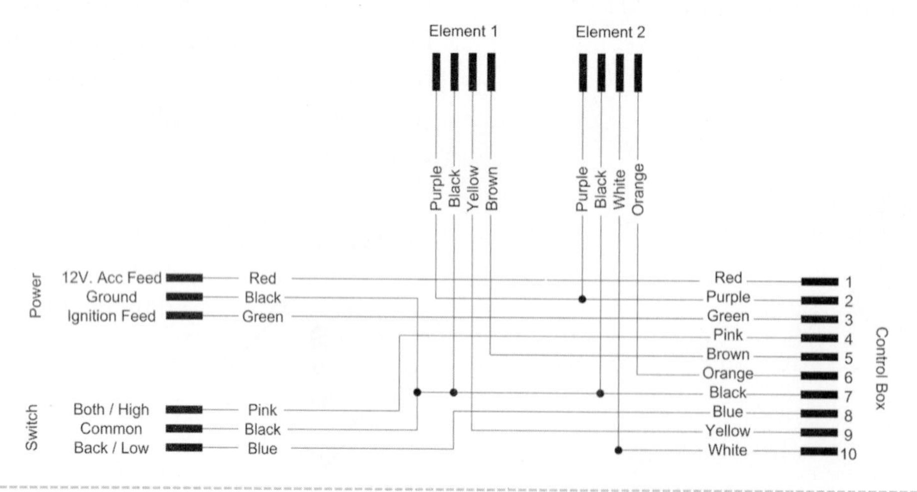

Figure 2-14 *Wiring diagram for seat heater kit*

- In Mode II, the user can select between a heated back or both regions heated.

By default, the kit comes supplied in the second of these modes. To put it in Mode I (low and high heat mode), the installer must cut away a wire loop shown in the wiring harness.

Choose a place to mount the switch that has sufficient room to allow the wires to pass behind. As with all the other projects in this book, route the cable harness to ensure the switch-mounting location has sufficient cable to reach the seat. If you need to modify the harness in any way or construct another one, a wiring diagram is given in Figure 2-14 on page 28.

One consideration is that the seat heater must be able to operate when the seat is in any position. Be sure that the wiring harness will not stretch when the seat is moved forward and backward.

Drill the hole for the switch and press the switch into the hole until a snap is heard; the switch is seated.

Installation is now complete!

Troubleshooting

Q. No heat can be felt when the device is switched on.

A. Check the accessory power feed. Check to be sure the fuse is operational and that all connections are sound. Check that a feed comes from the ignition when the key is inserted and in the correct position. Check the connection to ground and that any connections to the vehicle bodywork are not rusty or corroded. Make sure that the switch is operational.

If you suspect that the seat-heating element is at fault, you can take some readings using an ohmmeter.
The sensor wires should give a reading of between 19 and 22 K-Ohm at room temperature. The element itself (white wires) should give a reading of 5 Ohms or less.

Finally, check all connections on the wiring harness. Check to be sure connectors are correctly seated and cables are under no undue stress or strain.

Q. The heat felt is at a lower than desired level.

A. If the pad is covered by more than 1/4 inch of upholstery, the heat felt by the seat occupant will be reduced.

Project 12: Installing a Lumbar Support

A feature found on some high-end automobiles is an adjustable lumbar support. In its simplest form, it comprises an inflatable bladder in the seat back, which is pumped up by hand using a squeezable bulb with a nonreturn valve. To remove the lumbar support and deflate the bladder, the nonreturn valve is depressed and the weight of the passenger squeezes the air out of the bladder.

In this project, we are going to cover the installation of a high-end lumbar support. All movement of air is done by a small pump and compressor, resulting in the best possible user experience. As is the pattern with this book, I try to make everything electrically driven! No manual intervention is required.

A full kit of parts for the installation is available from Rostra. The kit is illustrated in Figure 2-15.

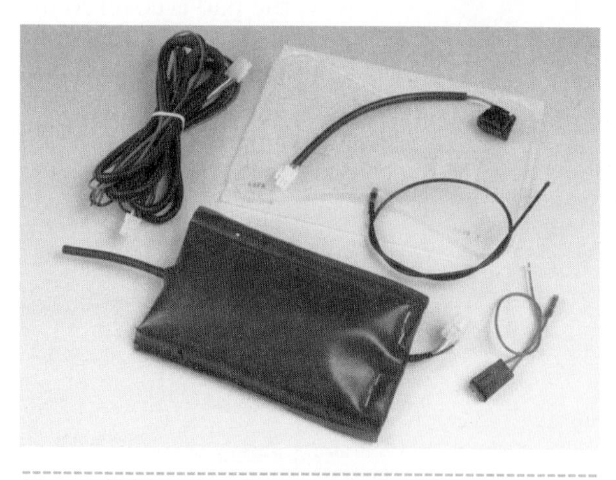

Figure 2-15 *Lumbar support kit (Courtesy Rostra Precision Controls)*

You Will Need

You Will Need

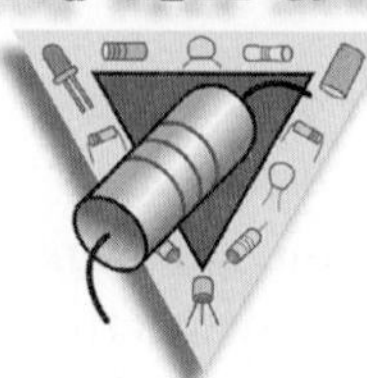

- Air bladder, choice of:
  - Standard bladder
  - Extended bladder
  - Small bladder
  - Front-mounted bladder
  - Pump and pouch assembly
  - Switch assembly
- Main wiring harness
- Fused harness

Tools

- Seat switch hole cutter, Rostra tool No. 250-1428 (optional)

Before starting the installation, make sure that you are familiar with all of the components and how to install the lumbar support.

You are going to need to remove the seat upholstery for the back of the seat. This procedure is covered in detail in Project 11, "Installing Heated Seats." As you are removing only the seat back, it may not be necessary to remove the whole seat to gain access. However, you may find it significantly easier if you do!

The type of bladder that you use will depend largely on the type of construction in the seat. Car seat construction falls roughly into two categories:

- If your seat is rigid, does not have any spring, or is made from a solid block of molded foam rubber or plastic, you will need to order the front-mounted bladder.
- If your seat is of a more conventional construction with springs and access to the inside of the seat, then you will use one of a range of bladders depending on the size of your seat.

Installation Instructions for a Front-Mounted Bladder

If you are installing a front-mounted bladder, you will need to secure the bladder so it sits in the lumbar region of the seat and does not move. There are a variety of ways that this can be done that vary with the type of seat construction and material, but a strong mounting using double-sided tape or cable ties is known to work.

Installation Instructions for a Rear-Mounted Bladder

You will need to gain access to the inside of the seat. In many cases, this can simply be done by lifting the springs and back of the seat clear. You then need to insert the bladder. The correct orientation is to have the inflatable bag side of the bladder facing the seat cushion, with the tube of the bladder pointing downward. The tube should then exit near one of the seat hinges and pass underneath the seat. The tube should pass between the hinges, not outside them. It can be secured with cable ties or zippy straps. The bladder can then be secured to the seat springs using a similar method.

Installation Instructions for a Seat Pouch or Pump

The pump must be mounted underneath the seat. Cable ties provide a secure method of installation.

The rubber hose can now be connected to the bladder. When connecting and routing the hoses, be sure of the following:

- The positioning of the tube does not conflict in any way with the seat runners or tracks.
- The tube is not kinked.
- No part of the bladder/tubes/pouch/pump is in contact with the metal springs, as they may impair the function of the device.

Installation of Wiring

A wiring diagram is given in Figure 2-16.

You will need to cut a hole for the switch at a suitable position on the vehicle. The dimensions are shown in the template in Figure 2.17.

This can be done with a tool purchased from Rostra. You will find that room temperature is optimum when cutting plastic used in automotive trim.

Snap the supplied bezel onto the switch and press the assembly into place.

The next job is routing the cables. If possible, this should be done behind trim panels and carpet to give a pleasing effect.

Switch Connector

Green
Red
Yellow
+12v

Pump Connector

Green
Yellow
Black

Figure 2-16 *Lumbar support schematic*

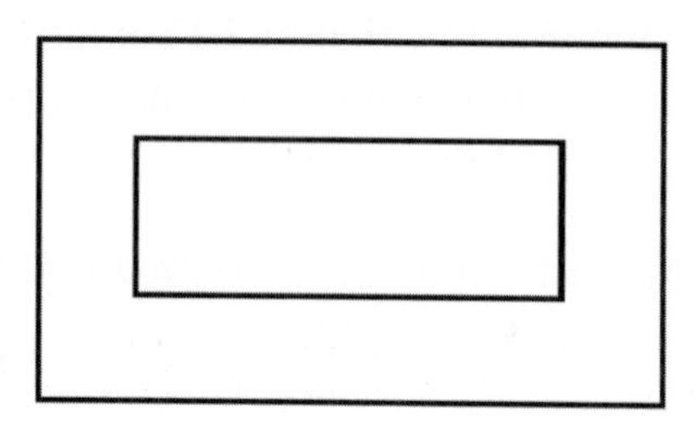

Figure 2-17 *Switch cutout*

Note

The supplied wiring loom is perfectly adequate for the job being discussed; however, if you have more complex requirements (for example, if you would like to control your lumbar support from a car PC, with a parallel port relay board), then the following should act as a good guide.

The red connection to the switch is the 12-volt power connection. Conversely, the black connection to the pouch/pump is a ground connection.

The control switch is a changeover switch that supplies power to either the pump (to pressurize the lumbar support) or the release valve (to depressurize the lumbar support). The green wire feeds the pump, while the yellow wire feeds the valve.

Troubleshooting

Q. I can hear the pump inflating and deflating, but the bladder is not filling with air.

A. Disconnect the pump/pouch assembly from the bladder. Now operate the switch and see whether the air is flowing from the pump. If the pump appears to be pushing air through the pipe, then the chances are the bladder is broken so replace it.

Chapter Three

Intelligent Functions

Intelligent functions are not intended to remove control from the driver but to augment and assist him or her in normal driving conditions. In exceptional circumstances, it may be necessary to override a car's intelligent functions. Therefore, the capability to override the car should always be included to allow the driver to intervene if necessary. When testing intelligent functions, first do so off road to check that the system works correctly.

Project 13: Installing an Automatic Headlamp Switch

A dangerous situation often arises when a driver forgets to turn on his or her lights in the dark. Unaware of the hazard they are causing, the driver relies on the light from streetlamps and other road users.

This circuit helps avoid such a situation. It is also failsafe in the respect that, should power to the relay be lost through circuit malfunction, the contacts of the relay will rest in the closed position—switching the lights on. As the ambient light level drops, the car's headlamps are automatically switched on. This circuit is surprisingly easy to accomplish with a minimum of components (see Figure 3-1 for constructional details of the circuit). It is better to drive around with lights on during the day than without lights at night!

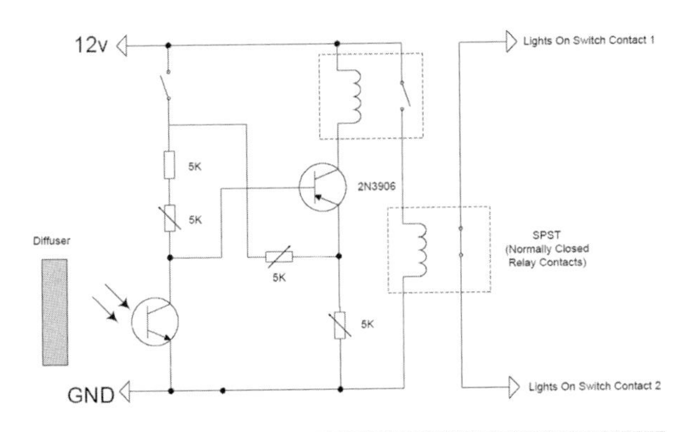

Figure 3-1 *Automatic headlamp switch schematic*

You Will Need

You Will Need

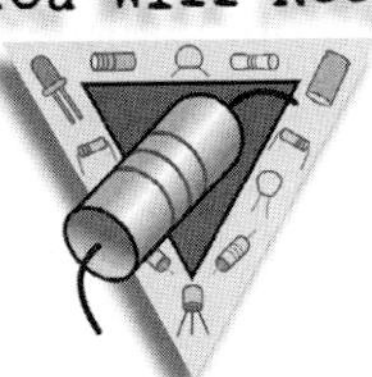

- 5K resistor
- 5K variable resistor (×3)
- NPN phototransistor
- 2N3906 PNP transistor
- 12 V single-pull single-throw (SPST) relay
- 12 V SPST relay, normally closed contacts (high current)
- SPST switch
- Photographic light meter diffuser
- Blue gelatin filter (optional)

Tools

Soldering iron

You will need to obtain a diffuser of some sort to ensure that the photodiode receives an average reading of the ambient light rather than a point-source reading. A suitable diffuser can be obtained from a good photography shop, as an accessory for a light meter. Failing this, a diffuser could be fabricated easily from some opaque or milky white plastic. You may find that the sensor is confused by the orange sodium glow of the streetlights. This can be remedied by a blue filter, which will block out orange light. As daylight contains a more balanced spectrum, it will be unaffected by the blue filter.

To ensure that the circuit is reliable in service and operates for a long period of time without malfunction, make sure the wires to the sensor are well insulated and that the sensor is protected against water ingress.

The sensor should be mounted on the roof of the car facing the sky. Try to position the sensor so that it is not affected by the light of either oncoming vehicles or following vehicles.

Do not attempt to adjust this circuit while the vehicle is in motion.

To calibrate this circuit, park it on a street with streetlights in the evening and adjust the potentiometers just past the point where they allow the relay to trigger. The circuit should automatically switch the lights on as the daylight fades. This circuit can be used with the automatic dip-beam lights.

Project 14: Installing Automatic Dip-Beam Lights

You Will Need

You Will Need

- 5K resistor
- 5K variable resistor (×3)
- NPN phototransistor
- 2N3906 PNP transistor
- 12 V SPST relay
- 12 V SPDT relay (high current)
- SPST switch
- Photographic light meter diffuser
- Blue gelatin filter (optional)

Tools

- Soldering iron

The circuit for the automatic dip-beam lights is essentially the same as the automatic headlamp switch. The difference lies in the mounting of the sensor and the type of relay used. The diffuser is omitted; instead the phototransistor must face forward, say, through the car's grill to allow it to capture the light from oncoming vehicles. A little experimentation may be required to gauge the optimum position (see Figure 3-2 for constructional details of the circuit).

Hint

To ensure that the circuit is reliable in service and operates for a long period of time without malfunction, make sure the wires to the sensor are well insulated and that the sensor is protected against water ingress.

Rather than getting a generalized reading of the ambient light, the sensor needs to point toward and take a light reading from the road in front to detect the presence of any oncoming vehicles.

To this end, the phototransistor should be mounted at the back of a dark tube. This gives the sensor more directionality.

Warning

Do not attempt to adjust this circuit while the vehicle is in motion.

To calibrate this circuit, park the car at one end of a short stretch of road facing the front of another vehicle. Adjust the potentiometers so that the relay triggers when the other vehicle's headlamps are on.

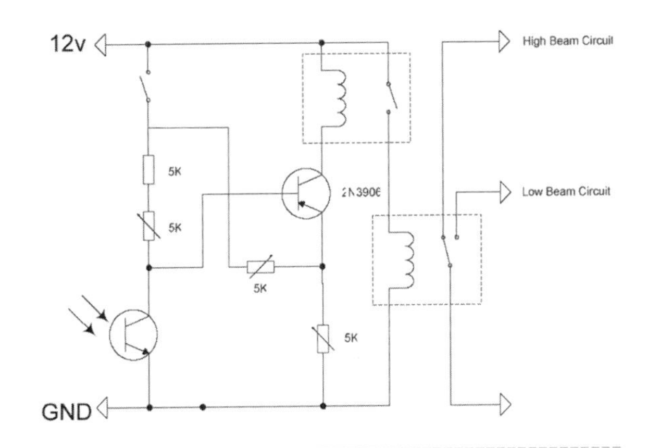

Figure 3-2 *Automatic dip beam lights schematic*

Project 15: Installing Simple Intelligent Wipers

For safe driving, it is imperative that the driver of the vehicle maintains good visibility. It is therefore of paramount importance that the vehicle's windshield wipers function effectively. Manual control of the vehicle's wipers may prove tiresome in adverse weather conditions. This simple circuit automatically switches the wipers on when droplets of water land on the sensor. The wipers will continue to operate until disabled.

Hint

The difference between the simple intelligent wipers and the advanced intelligent wipers is that the simple intelligent wipers are triggered by rain and will switch the wipers on only at a single speed. The advanced intelligent wipers, on the other hand, will automatically respond to varying weather loads imposed on the vehicle and adjust the wiper speed accordingly. It will also switch them off when they are no longer needed.

You Will Need

You Will Need

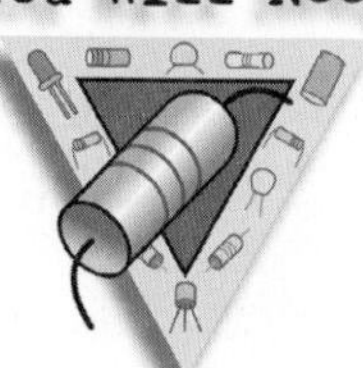

- Resistor 680R
- Resistor 1K
- Diode 1N1001
- Silicon-controlled rectifier C106B1
- 12 V SPST relay

Tools

- Soldering iron
- Side cutters

This automatic wiper sensor is very simple; it is designed to simply switch on the wipers if moisture is detected. The moisture sensor is a spiral of copper tracks exposed to the elements. In the event of bad weather, rain is detected by bridging the tracks, where solder or another conductor (in this case, water) crosses adjacent tracks and allows electricity to flow (see Figure 3-3).

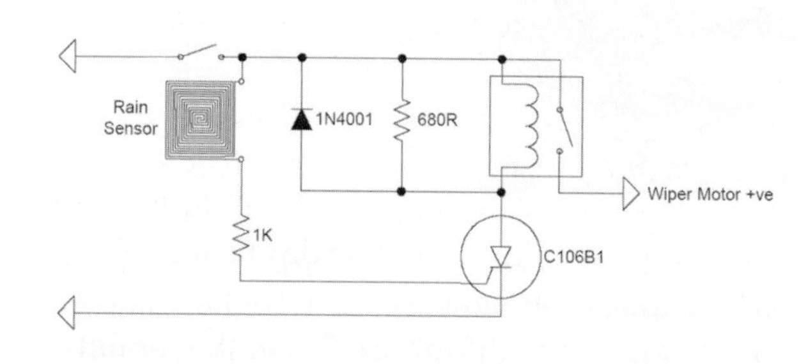

Figure 3-3 *Simple, intelligent wipers schematic*

Project 16: Constructing a Rain Sensor

You Will Need

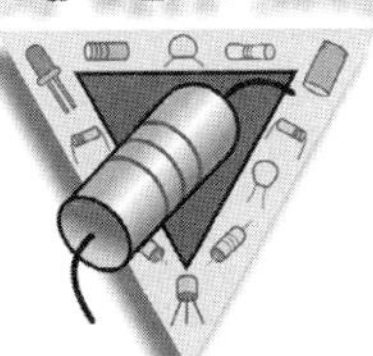

- Copper clad board (photo etching optional)
- Tin plating crystals
- Ferric chloride crystals

Tools

- Polypropylene tray
- Small printed circuit board (PCB) drill

In order to be able to construct the sensor, you will need to have the capability to etch small PCBs. Use the template supplied in Figure 3-4 as a guide. A variety of methods are available on the market for producing a PCB etch resist. The simplest is using an etch resist pen and a ruler. This should provide satisfactory results. Other methods include a photographic transfer by printing the design onto drafting film, placing it onto a sensitized board, and exposing it to *ultraviolet* (UV) light. A simple system that is very effective is the press-n-peel PCB transfer system. Photocopy or laser print the design onto a sheet of this material. Iron it onto a plain copper board using a household iron, peel the sheet away, and etch the board.

You will need to etch the board with ferric chloride crystals or something similar. Mix the crystals up according to the instructions on the packet. Be careful with the crystals as they are harmful and can irritate skin.

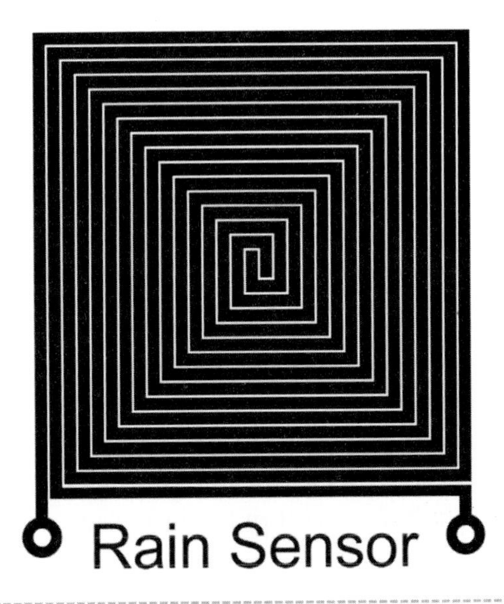

Figure 3-4 *Template for small PCB*

Once the board has been etched and dried, tin plating crystals can be used to deposit a coating of tin on the copper and make the board more durable.

As an alternative to etching a PCB, you could construct a sensor from a piece of stripboard. Connect every other track to one sensor wire and the remaining tracks to the other sensor. This is a less elegant solution but will be functional.

You may find that this type of sensor needs periodic replacement, as the copper corrodes. You can use a small piece of abrasive paper to restore the copper and remove verdigris.

Project 17: Installing Advanced Intelligent Wipers

Intelligent wipers come on when water falls on the windshield and then adapt to changing weather conditions. Wipers that adapt themselves to changing weather conditions are not only convenient, but they also improve safety.

Principle of Operation

A sensor with a light source and a receiver attaches to your windshield. The sensor is inclined at an angle toward the windshield. The light bounces off of the inside surface of the glass and is reflected back to the light sensor (see Figure 3-5). Electronic circuitry compares the value of the received light with the value stored in memory and comes to the conclusion that no rain is falling on the windshield, and therefore no action is needed.

When a little rain falls on the windshield, its optical properties are changed a little. Some of the light is diffused and lost from the windshield while some may be reflected (see Figure 3-6). The sensor receives a little less light now. The output of the sensor is fed to the comparator. The comparator stores the value in the memory and decides that a little rain is falling on the screen, and so an intermittent wipe would be appropriate. The wiper motor is pulsed intermittently.

Now a considerable amount of rain is falling on the windshield. Quite a lot of the light is being diffused so the sensor registers a relatively low reading. It is decided that a fast wipe is the only way to clear the screen, and so the wipers are switched on continuously (see Figure 3-7).

The sensor is placed centrally on the inside of the windshield in a region that is covered by the wipe of one wiper. As the wiper clears the windshield, the unit detects the lack of rain and will therefore switch off if the windshield has been successfully cleared (see Figure 3-8).

To install advanced intelligent wipers in your vehicle, you will need to obtain the following parts from Hella (see Appendix A for contact information).

Hella publishes a special applications list. This project applies to the majority of the vehicles on the road. However, if your vehicle is on the special applications list, you will need to follow a different installation procedure.

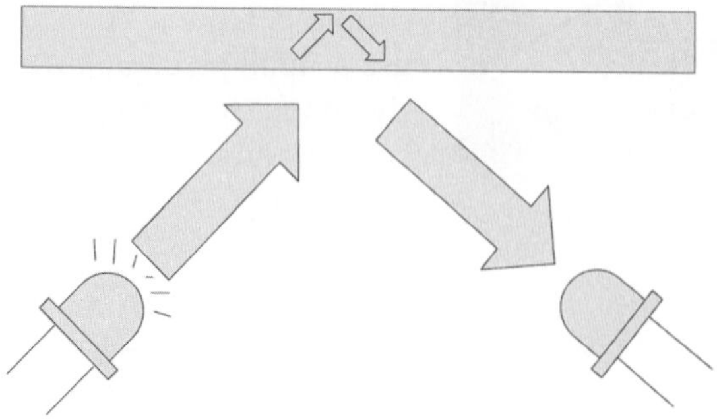

Figure 3-5 *Rain sensor with no rain falling on the windshield*

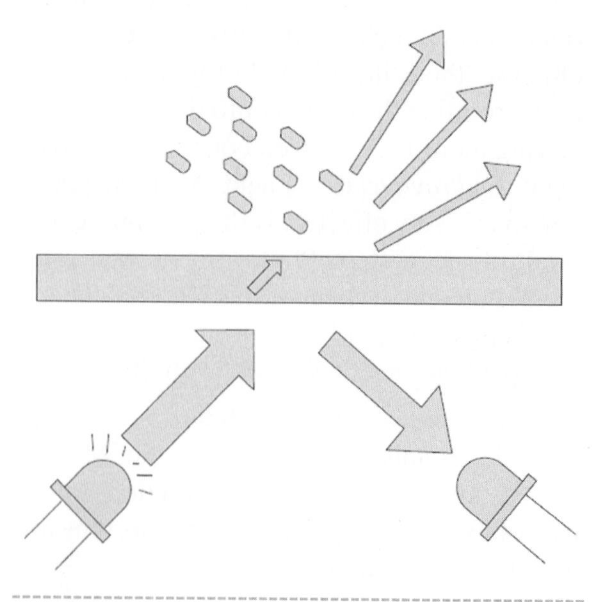

Figure 3-6 *Rain sensor during light rainfall*

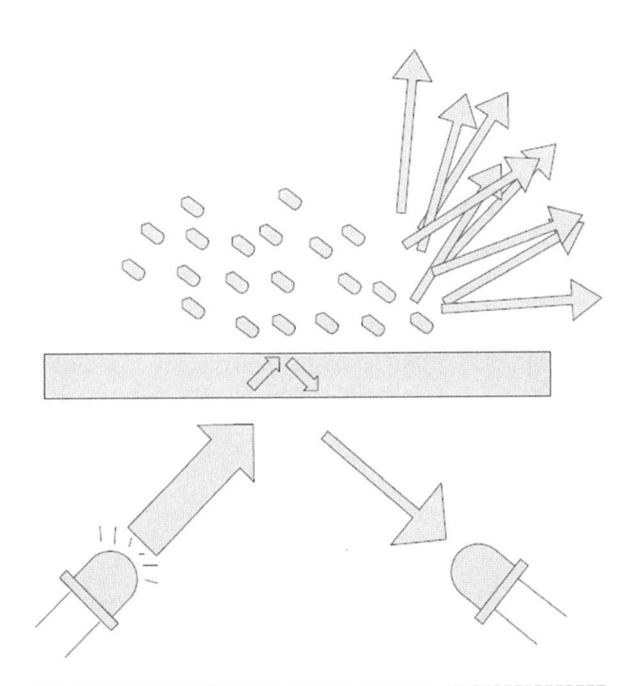

Figure 3-7 *Rain sensor during heavy rainfall*

Warning

Unfortunately, you will not be able to install this device in vehicles that have a heavily tinted front windshield or quick-clear windshields, which remove water by heating the glass (as this will cause the sensor adhesive to soften). Heated windshields may go by the brand name of Siglasol or Thermacontrol.

You Will Need

You Will Need

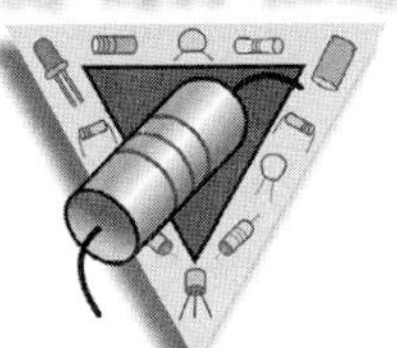

- Sensor unit: 9XB 733 069-001
- Electronics unit: 6PW 733 071-001
- Adapter: 5WB 007 977-001
- SPST latching switch (optional)

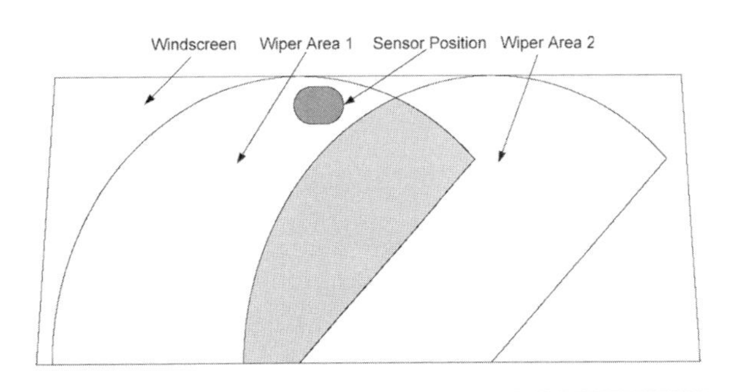

Figure 3-8 *Sensor placement on windshield and wiper areas*

Tools

- Side cutters
- Crimping tools

The package is sold under the brand name Hella Raintronic. The contents of the kit can be seen in Figure 3-9.

The kit comprises the connection cable (see left), the sensor housing (top right), the sensor electronics (second down), the replaceable sensor element (third down), and the relay module (bottom).

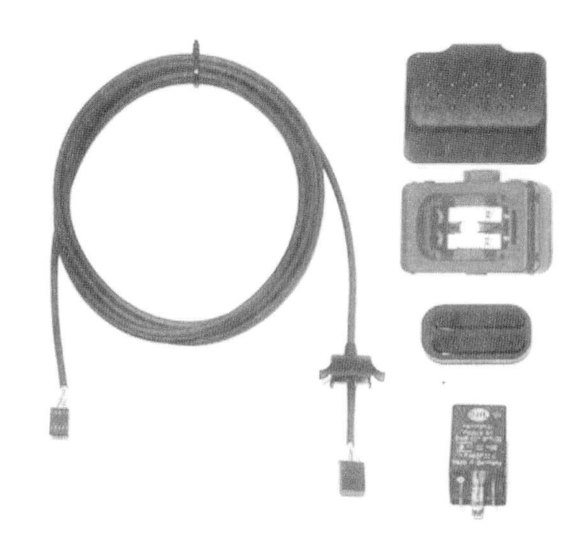

Figure 3-9 *Hella Raintronic Kit*

This is a sensor/relay package that controls your wipers according to the weather conditions. It will detect rain, snow, and dew, and it reacts faster than the human eye.

The Raintronic adjusts the wipers' stroke cycle depending on how much rain is falling on the windshield. It uses a sophisticated intelligent algorithm that "learns" and can adjust the sensor to compensate for wear on the wiper blades and windshield. The sensor will also detect that it is night and will adapt the wipe cycles accordingly for improved vision. The surface of the sensor is heated to ensure reliable operation in winter.

To begin installation of the intelligent wiper, first identify the relay that controls the wiper interval. This information may be found in your car's service manual. Failing that, it can be identified by touch, that is, by feeling which relay in the box vibrates when the wipers are functioning.

Remove the relay and compare the configuration of the spade terminals underneath to those on the replacement electronics unit (which resembles a relay with a socket atop). When the electronics relay module is removed, the wipers' interval function should no longer operate.

If the interval relay matches exactly, then it is a straight swap. Go on to the step called "Locating the sensor." If it is not a straight swap, further action is required.

Should the replacement electronics unit not match the relay, replace the relay and refer to Figure 3-10.

Locating the Sensor

The sensor must be located so that it is within range of the sweep of one wiper (refer to Figure 3-8).

The wiper must overlap the sensor by at least 2 cm (see Figure 3-8). The sensor must be stuck to the inside of the windshield. If there is any sun stripe, shading, or other discoloration at the top of the windshield, the sensor must be mounted well clear of this.

CAUTION

Be sure that when you mount the sensor, you leave enough room for your rearview mirror to rotate correctly through its full range of motion and that you allow sufficient space for the cable and housing of the sensor.

The windshield should be cleaned thoroughly using a lint-free cloth to prevent any fibers from sticking to it. No solvent or detergent should be used. The glass should be very dry before applying the sensor. Do not clean the glass with any cleaning fluid for at least 24 hours after applying the sensor, as the adhesive needs time to dry.

The adhesive should be heated using a radiant heater, hair dryer, hot plate, or other heat source until the surface is about 170 to 190° Fahrenheit (80 to 90° Celsius). The sensor comes covered with an adhesive film that must be removed before affixing the sensor.

Apply the sensor so that the longest edge touches the windshield. Now roll the sensor downward so that the adhesive is gradually applied. This rolling motion helps prevent any air bubbles from forming.

Check from the outside of the vehicle that no large air bubbles have formed (these would prevent the sensor from functioning effectively). If any air bubbles do become trapped between the sensor and the windshield, you will need to purchase a new sensor unit to clip onto the electronics unit. Heating the windshield gently will allow you to remove the old sensor safely. A twisting motion can be used to remove the sensor unit, but be careful not to damage the electronics in the process.

A chemical cleaner can then be used to remove any traces of glue on the windshield.

Warning Be sure to let the windshield dry thoroughly before applying the new sensor.

To remove the sensor unit from its housing, some small tabs must be slid out of the way of the electronics unit using a small screwdriver.

Warning If you are mounting this unit in a large SUV or commercial vehicle, you may find that the wipers do not extend fully to the top of the windshield. This would result in a long dangling wire, which would be distracting for the driver as well as not being mechanically sound. In these instances, it is possible to mount the sensor on a flat surface, in the bottom right-hand side of the windshield.

Electrical Installation

A pinout of the Raintronic Control relay can be found in Figure 3-10.

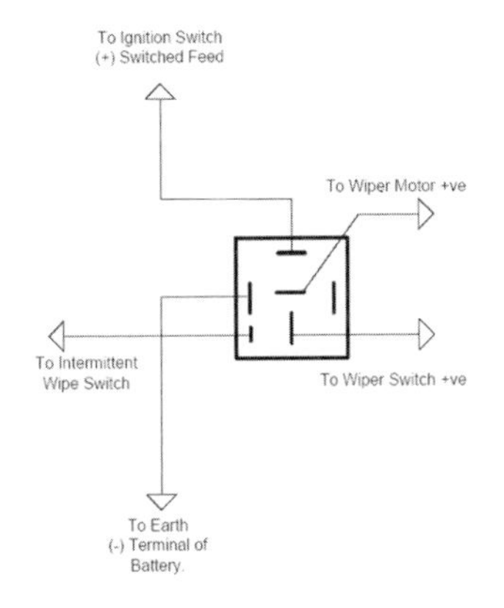

Figure 3-10 *Raintronic Wiring*

Contact 15 will connect to the ground, and contact 31 connects to an ignition-switched live terminal.

The remaining connectors are I 53M and 53S. You will need to locate the wiring loom between the wiper switch and the windshield wiper motors. The wire that powers the wiper motor must be cut; 53S connects to the portion of wire that leads to the motor, while 53M must go to the vehicle wiper switch. The I terminal is connected to the intermittent wipe switch (with the connection to the motor being cut) or is connected to a separate switch that is connected to the ignition-switched live terminal.

There is a six-contact plug that will connect to the electronics sensor unit. It comes with a small plastic shroud that will snap into the plastic casing for cosmetic purposes. The cable can then be run behind your headlining and down one of the A pillars. If you are in a left-hand drive car, the chances are that your fuses and relays will be somewhere in the driver's compartment to the left-hand side of the vehicle. Conversely, if you are in a right-hand drive car, it is likely that the fuse and relay box will be on the right-hand side of your vehicle. It is likely that this will influence your choice of A pillar!

You may find that removing your sun visors and door seals will help when trying to conceal the cable, as these items of trim often secure the underlying headlining and panel trim.

Troubleshooting

If the wipers do not operate when the interval switch is turned on (or the optional switch where fitted), then the first course of action must be to check for loose plugs and connections. Next a quick look at the fuse box is in order; check that everything is operational.

Check that the plug from the sensor unit is installed firmly into the slot at the top of the new relay. And finally, check that the ignition is switched on. Things function much more impressively when supplied with power!

To test the system, the first basic test you should perform is to switch the interval (or optional switch where fitted) on and look for the telltale single wipe that acknowledges the unit is responding.

Switch to your first wiper-level setting if there is still no response. You may find wires 53S and 53M are connected the wrong way around. The fuse will have blown as a result.

Switch between the first wiper-level setting and the no-wipe position. If there is no response or the blades are stuck and not parked, you should check to be sure the relay is plugged in securely and in the correct socket.

If, on the other hand, you find that the wipers will not stop wiping when the windshield is dry and the wiper switch is in the interval position, check the obvious. Be sure connections are secure and there are no loose plugs.

Then reset the device. If the wipers continue to wipe, one of two things may be the matter:

- Air bubbles may have gotten under the sensor, preventing it from operating correctly.
- The windscreen creates a heat-insulating effect due to the infrared radiation from the unit.

Resetting the Raintronic

Take the following steps to reset the Raintronic wiper:

1. Turn the ignition on.
2. Turn the interval switch (or optional switch) on and off twice during the wiper stroke.
3. Leave the ignition on for an additional 5 seconds.

Reinitializing the Raintronic

Take the following steps to reinitialize the Raintronic:

1. Turn the ignition on.
2. Turn the interval switch (or optional switch) on and off twice during the wiper stroke.
3. Switch off the ignition within 2 seconds of the wiper finishing its first cycle.

Advanced Intelligent Wipers in Action!

Calibrating the System

Now it comes to the nitty-gritty of actually using the advanced intelligent wipers! Before you can use them, you must go through a simple calibration procedure. Sitting in the driver's seat, with the ignition switch off, turn your wipers to their interval setting. Now switch on the ignition. This will put the wipers into teach mode. The sensor will now adapt to how clear your windshield is. This is followed by a wipe of the wipers for confirmation. The next piece of programming is done in the rain. This may take up to 15 minutes, and it is not advisable to do this while driving!

The second piece of learning is done with the wipers in interval mode. The system will optimize itself by evaluating the wipers' system and the position of the blades. The unit will perform several wipes and see how the system performs, adjusting settings for maximum performance.

Using the System

To use the system, simply put your wipers into interval mode and let them run. A single stroke of the wipers will confirm that the system has been initialized. The unit remains on until you switch it off. Where you have fitted a separate switch, this will take the place of the Raintronic unit.

Every time you replace your wiper blades or change something about the wiper system, you should reprogram your Raintronic to ensure optimum performance.

Project 18: Installing Cruise Control

Cruise control is a feature that is fitted to many luxury vehicles. Ideal for people who need to travel long distances, it allows the user to maintain any given speed and increase it or decrease as traffic conditions change, all while resting his or her feet away from the pedals! By holding the vehicle at a steady speed, the miles per gallon and therefore savings should increase.

During the 1950s, a simple cruise control appeared in the form of the hand throttle. This usually took the form of a steering-wheel-mounted lever that could be set in any position, locking the throttle cable at that point and allowing the feet to be taken off the pedals. This system had a number of disadvantages despite being a simple system.

As road conditions change, so will engine load. This system makes no allowances for hills or varying engine loads. As a result, the car will speed up going down hills or slow down going up hills.

Modern electronics can put a more sophisticated cruise control within the grasp of the weekend hobbyist.

A car's speed is regulated by a throttle actuator. If the actuator is electrical, it will be powered by a stepper motor, and if mechanical, it will be driven by the vacuum system in the intake manifold.

In this project, we will cover the installation of an electromechanical system, which seems to be the most prevalent in aftermarket kits. The particular kit I will be discussing is the GlobalCruise, which is available from Rostra Precision Controls. This project is written to be generic in its scope and so should cover pretty much any aftermarket kit, but the Rostra GlobalCruise is my cruise control kit of choice.

The GlobalCruise drives the throttle cable by means of an electric motor. This installation is simpler than one for a vacuum-actuated kit, as you do not need to take a tap from the intake vacuum pipe. Should you wish to install a vacuum-actuated kit, you can find one through Rostra as well. The difference with a vacuum-driven kit is that the throttle cable is not connected to a motor; instead it is connected to a diaphragm in a sealed chamber. This diaphragm flexes with differences in pressure created by the engine vacuum. Electronically actuated valves control the pressure to the diaphragm and therefore control the throttle.

In an electronic cruise control system, the driver is presented with a number of push button controls:

- Set
- Accelerate
- Decelerate

Safety and the Cruise Control

The system is only as good as the input it receives from its sensors. A number of wires are "optional" in the installation procedure. I recommend you connect them all!

The cruise control can be a very effective safety tool. For example, under aquaplane conditions, the cruise control will quickly detect the wheels spinning, consequently reducing engine speed and keeping the driver in control of the vehicle.

Remember when driving, the cruise control unit will attempt to keep the vehicle's speed at the set value. It cannot anticipate road conditions. Therefore, if the speed is high and you approach a corner, the unit will attempt to keep the speed set.

Cruise control is most suited to driving at a steady speed on straight roads such as on the highway. It is not recommended for urban driving, heavy traffic, winding country roads, or loose road surfaces.

You Will Need

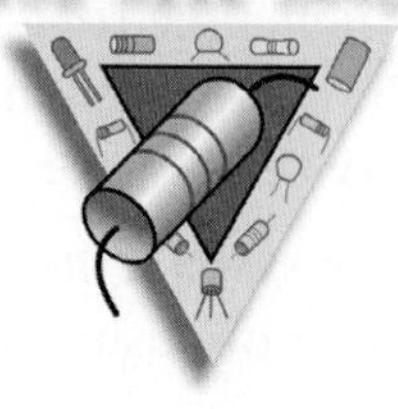

- Cruise module
- Cruise harness
- Cruise cable
- Module bracket
- Cable bracket
- Convoluted tubing 58 inch (145cm)
- Hardware package
- Module bolt
- Self-threading bolt (M6 × 19)
- Bead chain
- Bead chain connector
- Connector cover
- Loop cable (67mm)
- Loop cable (81mm)
- Three-bead connector
- Eyelet connector
- Tie strap (102mm)
- Tie strap (190mm)
- Tube Clamp (10mm)
- Flag nut (threaded tube clamp)
- M5 bolt (M5-.8×10)
- M5 bolt (M5-.8×20)
- M5 nut
- Locknut (nylon insert M5-.8)
- Lock washer nut (1/4 - 20)
- Plain washer
- Snap-in adapter
- Cotter pin (2mm × 16mm)
- Sealing putty (or 1-inch grommet)

The contents of a cruise control kit are illustrated in Figure 3-11.

Depending on your application and type of vehicle, a range of supplemental parts are available from Rostra that may aid the installation of your cruise control kit.

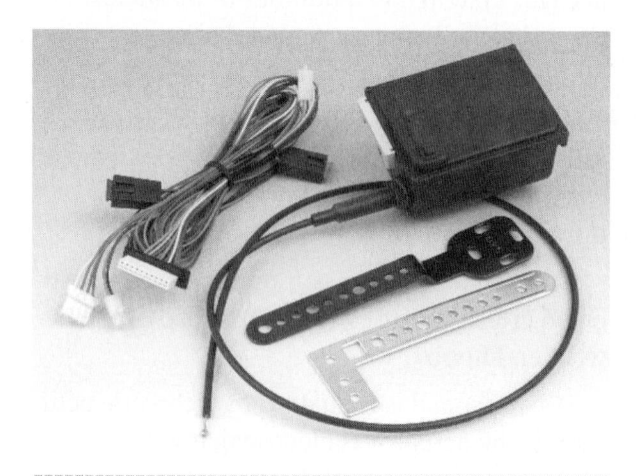

Figure 3-11 *Cruise control kit*

Tools

- Permanent marker
- Selection of screwdrivers
- Selection of wrenches and sockets
- Side cutters
- Soldering iron

This project is intended as a guide. Be sure to follow the instructions and manufacturer's data supplied with your cruise kit and adhere to their warnings. This project was written based on the manufacturer's data and actual installation of a GlobalCruise kit.

To begin, find a position to mount the actuator. Ideally this should be low down in the engine bay, making sure that the throttle linkage cable will reach the throttle linkage. Try to locate the unit away from any components, such as a radiator, which could cause variations in temperature.

Hold the unit in place temporarily with masking tape. You may need to adjust the unit later if cables do not reach. You will secure it later in the project when you are confident all components will reach.

The module bracket joins to the module with the aid of two bolts. Do not overtighten them. It is possible to shape the bracket for different applications.

There are a number of places where you should not mount the module:

- Under the fender (wing)
- Under the vehicle
- Directly to the engine
- With the cable pointed down
- Near sharp, hot, or moving objects (fan, radiator, and so on)
- Near the ignition coil
- In the passenger compartment
- Where it will obstruct access to other vehicle components

Drill holes in the bodywork and affix the unit to it using the self-threading bolts. Ensure that the programming switches are easily accessible or set the switches before mounting.

In the GlobalCruise kit, the throttle cable can travel 41mm. This will vary from kit to kit. With the car's engine switched off, you will need to measure the difference between the throttle cable position at idle and when wide open. Make a mark on the cable with a permanent marker. Align a ruler with the mark and get a friend to depress the accelerator fully. Measure the distance of the cable travel.

The instructions that you follow will differ slightly depending on the distance of travel.

Do this first if your cable travels less than 41mm. If the distance the cable moves is less than 41mm, you will need to add some length to the cruise cable in order to provide a little extra slack. Slack in the cable is defined as the amount that the cruise control cable moves before the throttle starts to move. By introducing slack, you allow a little movement before the throttle starts to operate.

This slack is introduced by using a bead chain; the kit should come with a number of links. These connect to a bead chain on the end of the throttle linkage cable, much as you would find

at home connecting your bath or sink plug to the bath or sink. A connector cover slides over the throttle loop cable and the cruise cable. A bead chain connector can then be installed on each. Beads can then be added to extend the cable. Each bead adds 0.28 inches (7mm) of slack. You need to add the number of beads that adds an appropriate amount of slack to enable the travel to be 41mm (this is the amount you measured earlier in the project). The bead connectors are lightly crimped to ensure the beads stay in place. The connector covers can then be placed over the joint to ensure that they are protected.
See Figure 3-12 for an illustration of the bead chain in place.

The throttle linkage cable is a belden cable, that is to say, a metal cable that freely runs inside a protective sheath. Because the sheath is secured, the inner cable can be used to convey the movement of a motor inside the throttle actuator to the throttle shaft of the carburetor or throttle body.

It can be attached to the throttle using a variety of methods depending on the application. In the Rostra kit, five different types of linkages are supplied. You will need to evaluate your requirements and identify the most suitable:

- Ford-type throttle linkage (for use with Ford vehicles)
- GM/Chrysler-type throttle linkage, three-bead connector (for use with GM, Chrysler, Vauxhall, and Daewoo cars)
- Loop cable pulley assembly
- T-bar adapter pulley assembly
- Connection to pedal

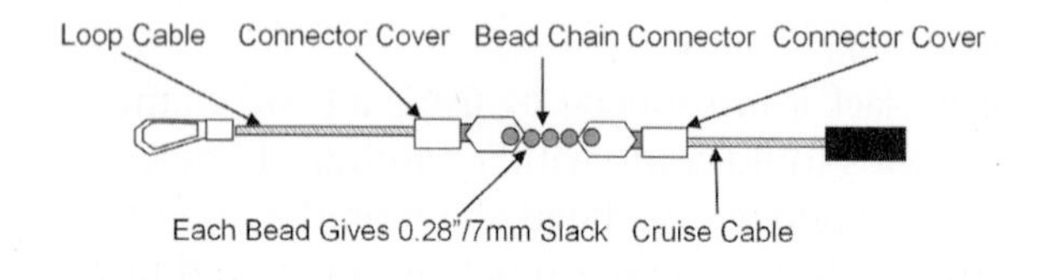

Figure 3-12 *A bead chain in place*

The cruise cable needs to be anchored securely so that the outer sheath of the belden cable is stationary while the inner section is free to move. The kit supplies several options for this. The first is a snap-in adapter, which is universal and will fit in many vehicles. This option requires a thread to be made on the end of the cruise control cable, which can be easily accomplished using the supplied lock washer nut.

The other supplied methods for securing the cable are a flag nut, which can also be used in many applications, and a "blank anchor," which is primarily used with General Motors vehicles.

The cable should be secured using P clips to the vehicle bodywork, avoiding the following:

- Any sharp kinks in the cable
- Any moving components that could chafe or damage the cable
- Any hot components that could melt the cable

When securing the P clips, do not overtighten them, as this may cause the actuator cable to jam.

Once all of the components are secure, the electrical installation is complete.

Electrical Installation

The electrical installation serves to connect the main cruise control module with the vehicle's ancillary systems and also to the cruise control panel.

There are a wide variety of cruise panels available from Rostra. If you do not like a button-press type of control, it is possible to buy a column-mounted control that attaches to your steering wheel binnacle. Figure 3-13 is an example of the type of column-mounted control available.

Figure 3-13 *Column-mounted control*

The wiring diagram for the Rostra Global-Cruise module is shown in Figure 3-14. A summary of all the connections follows.

Black: Ground

The ground wire must be connected to a good clean ground. If possible, tap ground from a suitable cable. But in the event that this is not available, tap ground from a good clean piece of metal with the paintwork uncovered. A little grease should prevent this connection from corroding. Under no circumstances should you use the engine or any bracketry as a grounding point.

Brown: Accessory Live Feed

This wire should be connected to ignition switch position one, that is, live when the key is turned one click away from off, but still operable when the engine is not running.

Red: Brake Switch "Hot" Side (Positive)

This wire should be connected to the side of the brake switch that is supplied with power at all times, even when the pedal is not depressed. It forms part of a failsafe circuit and is safety critical.

Violet/Purple: Brake Switch "Cold" Side (Negative)

This wire should be connected to the side of the brake switch that is live when the pedal is depressed. Again this connection is safety critical, and it is imperative that a sound connection is made.

Dark Blue: Tachometer Feed

This wire is used to sense the speed at which the engine is turning and prevent damage from overrevving. This connection is optional, but highly recommended as it provides additional safety. Where a clutch-disengage switch is fitted, it is recommended that this wire is grounded to prevent erroneous signals. This wire is also particularly useful in engines with an automatic transmission to prevent overrevving.

Grey: Vehicle Speed Sensor (VSS) Wire

Some vehicles with an electronic speedometer will have a vehicle speed sensor wire. Consult with a service and repair manual to see if your vehicle comes with this feature. If your vehicle does not have a vehicle speed sensor output, then you will need to purchase an additional sensor from Rostra that senses rotation magnetically.

Rostra provides a fax-back service for information on common models. Send a fax to (910) 610-4191 detailing your vehicle make, model, and year, and Rostra will endeavor to help you. Or call Rostra's Technical Service Department at (910) 277-1828.

Light Green:

Neutral Safety Sensor (NSS) The wire is optional and is useful for vehicles with an automatic transmission. It prevents the engine from overrevving when the gear selector is in the neutral position. It is not essential that this

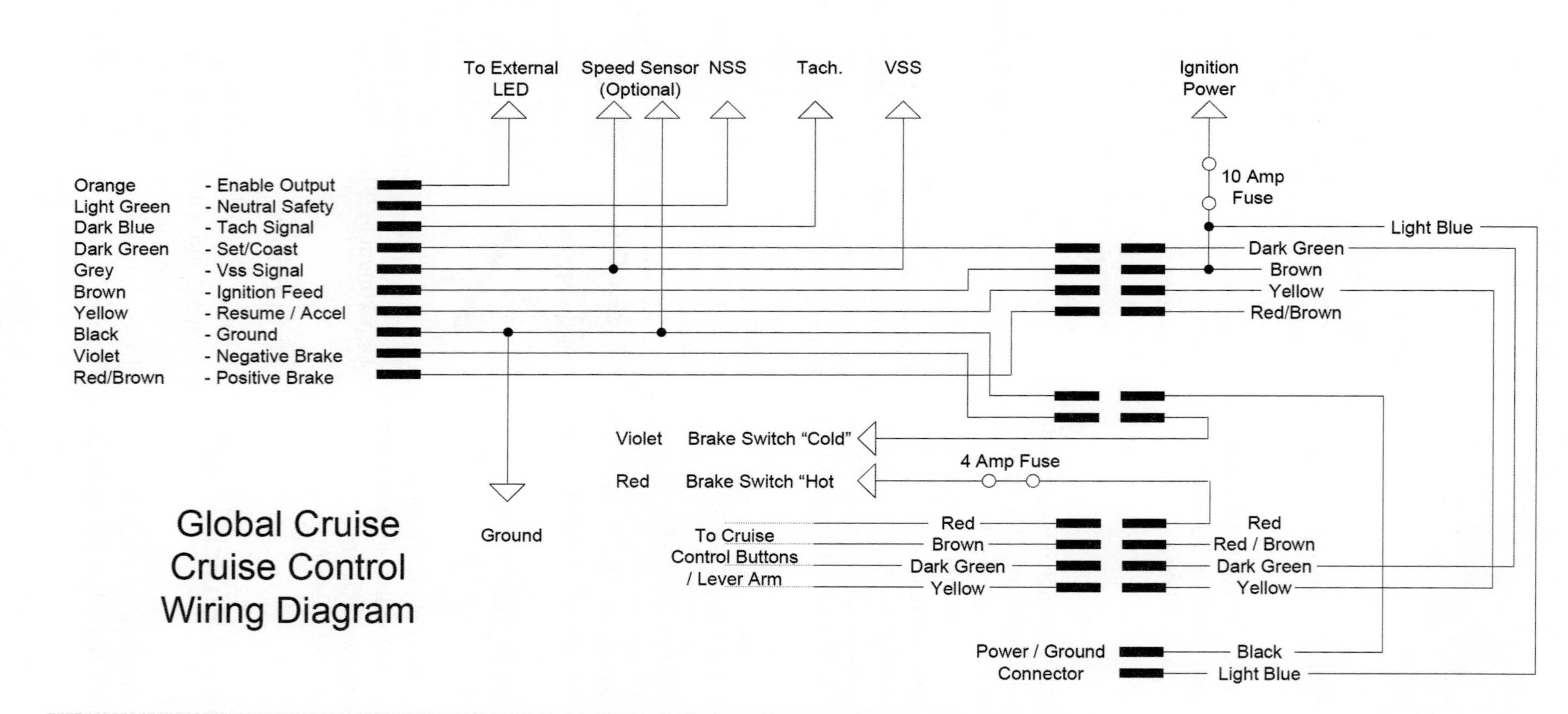

Figure 3-14 *Rostra Global Cruise diagram*

wire is connected, but the safety of the installation is greatly enhanced, so connection is highly recommended.

Orange: Enable Output

This output can be used to drive a signal lamp for a dashboard indicator. The signal wire will be driven low when the cruise module is enabled. This means that the other wire leading to the lamp must be connected to a 12-volt source.

Auxiliary Speed Sensor Input

When the gray VSS wire is not used, the optional kit must connect to the auxiliary speed sensor input. The gray wire is left terminated and unconnected.

Four-Way Switch Connection

All Rostra cruise control switches, including column-mounted ones, have a switch that will mate with this connection in the harness. It is simply a matter of connecting the switch and installing it in place.

Two-Way Switch Connection

This connection is used to provide additional power to switch modules that have illumination.

The wiring loom comes with a male- and female-mounted connection inserted in the place where the loom must pass through the bulkhead. Disconnecting this pair before installation and connecting them through the bulkhead will make installation considerably easier.

You will also need to configure the DIP switch in accordance with your vehicle particulars. The DIP switch is located on the cruise control module. It looks something like Figure 3-15.

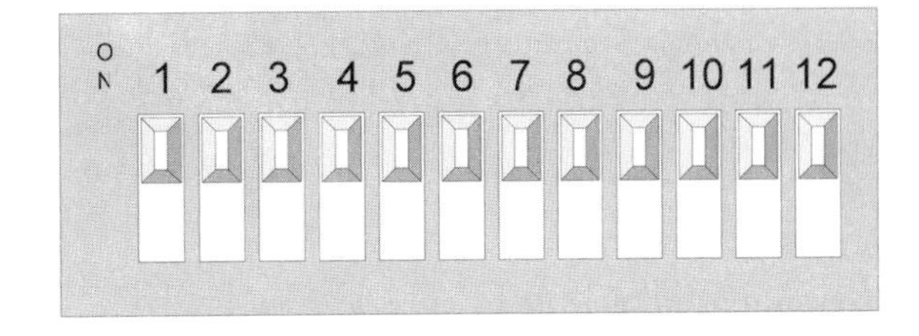

Figure 3-15 *DIP switch on a cruise control*

Refer to Table 3-1 for guidance on how to set the switches for your particular vehicle.

Troubleshooting

Rostra has thoughtfully included a diagnostic self-test LED close to the DIP switches used for programming the cruise module. This can be used to diagnose many common faults.

Entering the Diagnostic Mode

From the driver's position, switch the cruise control unit off, followed by the ignition key.

Determine whether you have a closed- or open-loop control switch. You will be able to identify this from the Rostra part number or by looking at the data supplied with your switch. The instructions will differ slightly.

Closed-Loop Control Switch

If you have a closed-loop control switch, do the following to enter diagnostic mode. Hold the resume/accelerate button down while you turn the ignition to position 1. Take your finger off the resume/accelerate button.

Programming Functions	*1*	*2*	*3*	*4*	*5*	*6*	*7*	*8*	*9*	*10*	*11*	*12*
Gain (Sensitivity)												
Extra Low	off	off										
Low	on	off										
Mid	off	on										
High	on	on										
Pulses per mile (Pulses per kilometer)												
2,000 (1,250)			off	off	off	off						
4,000 (2,500)			on	off	off	off						
6,000 (3,700)			off	on	off	off						
8,000 (5,000)			on	on	off	off						
10,000 (6,250)			off	off	on	off						
12,000 (7,500)			on	off	on	off						
18,000 (11,200)			off	on	on	off						
24,000 (15,000)			on	on	on	off						
3,200 (2,000)			off	off	off	on						
6,400 (4,000)			on	off	off	on						
9,650 (6,000)			off	on	off	on						
12,870 (8,000)			on	on	off	on						
16,090 (10,000)			off	off	on	on						
19,300 (21,000)			on	off	on	on						
28,960 (18,000)			off	on	on	on						
38,600 (24,000)			on	on	on	on						
Engine/Setup Timer												
8 Cylinder/Low						off	off	off				
4 Cylinder/Low						on	off	off				
6 Cylinder/Low						off	on	off				
6 Cylinder/Extra High						on	on	off				
8 Cylinder/High						off	off	on				
4 Cylinder/High						on	off	on				
6 Cylinder/High						off	on	on				
4 Cylinder/Extra High						on	on	on				

Programming Functions	*1*	*2*	*3*	*4*	*5*	*6*	*7*	*8*	*9*	*10*	*11*	*12*
VSS Source												
Sine Wave Input										off		
Square Wave Input										on		
Transmission												
Manual											off	
Automatic											on	
Control Switch												
Open Circuit												off
Closed Circuit												on

Open-Loop Control Switch

If you have an open-loop control switch, do the following to enter diagnostic mode. Turn the ignition to the on position, hold the resume/accelerate button down, and switch the unit on. Now take your finger off the resume/accelerate button.

The diagnostic LED will not be lit at this point. You have entered diagnostic mode.

Testing the Set/Coast Button

Press the button. The diagnostic LED should light for all the time that the button is held down.

In the event of the lamp not lighting, check the position of diagnostic switch 12. It will need to be set in accordance with Table 3-1 depending on whether you are using an open- or closed-loop switch.

Testing the Resume/Accelerate Button

Press the button. The diagnostic LED should light for all the time that the button is held down.

Testing the Brake Switch

You can perform this test using either a second person or a heavy weight to depress the brake pedal. When the brake pedal is depressed, the diagnostic LED should be illuminated. Conversely, when the pedal is in the resting position, the lamp should be off.

If this does not work, then the red wire, which leads to the hot or permanently live side of the brake switch, may be faulty.

Testing the Vehicles Own Speed Sensor (Where Connected)

To test the vehicle's speed sensor, release the handbrake and roll the vehicle forward and backward at a steady speed for a few yards. The light should flash at a constant rate proportional to the speed at which the vehicle is moving.

If this is not happening, check to be sure programming switch 10 is set to the correct position. It should be in square wave input mode.

If this is not the case, the vehicle speed sensor or the connection leading to it may be faulty.

Another method of testing the vehicle speed sensor is to jack up the drive wheels of the vehicle and with the vehicle supported on axle stands, spin the driven wheel vigorously. (If a *limited slip differential* [LSD] is fitted, you will need a partner to simultaneously spin the other driven wheel.)

Testing the Auxiliary Speed Sensor (Where Fitted)

Go to the wheel that the auxiliary speed sensor is fitted to. Jack up the vehicle and spin the wheel at a speed that would correspond to a road speed of about 3 miles an hour or faster. The diagnostic LED should repeatedly flash at a speed proportional to the speed of spin.

Check that programming switch 10 is set to off so that it accepts a sine wave input.

This completes the self-diagnostic testing procedure. If none of the above works, try the following steps.

Testing the Tachometer Feed

Warning

Make sure that the handbrake is on and the car is in neutral!

Follow the previous instructions for entering self-diagnostic mode, but rather than turning the ignition key to position one, continue to turn the key and start the engine.

The LED on the cruise unit should be flashing steadily as the engine idles steadily. Observe how the speed of flash changes as the engine speed is varied. Lightly depress the accelerator. The light should speed up accordingly. This confirms that the tachometer feed is connected correctly.

If this does not work, do the following:

- Check that a sound connection exists with the tachometer feed.
- Check that the tachometer feed is connected to the correct place.

Using the Cruise Control

Driving with cruise control will take a little acclimatization. However, once mastered, it will take the strain out of long-haul driving.

To switch the unit on is simply a matter of sliding the switch to the on position. If fitted with an indicator, the LED will light.

The system will engage only at speeds above 30 mph. Engaging the cruise control is a simple matter of pressing the set/coast button and removing your foot from the accelerator pedal.

Your speed will now be maintained to within approximately 2 mph of your set speed.

You can increase the speed of the vehicle by pressing the accelerator pedal. However, releasing the pedal will allow the vehicle to return to the set speed.

To slow the vehicle down, hold the switch toward the coast position. You should feel the vehicle slow. If you now release the switch, the vehicle will maintain the new set speed.

To accelerate, push the switch in the opposite direction toward the accelerate position. You will feel the vehicle accelerating. Releasing the switch will make the vehicle maintain the new set speed.

If you want to increase the speed in small increments, you can tap the cruise control switch up or down momentarily. This should increase or decrease your speed by about 2 to 3 mph per tap.

If you want to disengage your cruise control, simply tap your brake pedal. This will retain the set speed in memory. Pressing the resume button will allow the vehicle to accelerate back to this speed.

If you switch the cruise control unit to off, the cruise control's memory will be lost and a new set speed will have to be programmed in order to be able to resume to that speed.

Warning

You should not use the cruise control under adverse weather or traffic conditions.

If your car has a manual transmission, do not change gear without first disengaging the cruise control as you will cause the engine to overrev.

Looking to the Future

Two companies, Delphi and TRW, are currently developing the next generation of cruise control systems that will be appearing on a vehicle near you shortly.

Rather than simply maintaining a set speed, adaptive cruise controls evaluate the distance between the vehicle and any obstructions in front of it to provide a safer system. Adaptive cruise controls can maintain a set speed until a vehicle in front slows and encroaches on a set distance. They can follow the vehicle in front at a preset safe distance, or they will accelerate until vehicles in front are detected.

Adaptive cruise controls will be useful in heavy traffic conditions where present cruise controls are useless.

In addition to the components of a standard cruise control system, adaptive cruise controls have a radar system mounted behind the grill of the vehicle.

The radar system comprises a radar headway sensor, digital signal processor (which interprets conditions and processes the information from the headway sensor), and a longitudinal controller. In addition to throttle control, adaptive cruise control systems must have control of the braking system in order to slow a vehicle safely.

Adaptive cruise systems will be able to detect obstructions up to 500 yards ahead and will operate in the range of 15 to 110 miles an hour. In addition to cruise control, these systems will be able to warn the driver of an impending collision, allowing them to take evasive action.

Chapter Four

Security and Safety Projects

Vehicle safety has improved greatly as a result of electronics features fitted to modern cars. With advances in optoelectronics, LEDs can now replace incandescent bulbs, offering greater efficacy, quicker response time, and brighter light output. Many innovations, such as airbags, are possible only because of sophisticated microelectronics. Unfortunately, with all of the modern microelectronics packed into today's car, some in-car entertainment systems cost thousands of dollars, providing the opportunist thief with a more desirable target than ever before.

Fortunately, electronics again comes to the rescue, providing sophisticated security and tracking devices that can aid the police in tracking a vehicle's movement. In this chapter, we will be looking at some simple circuits, modules, and devices that will greatly enhance the safety and security of your vehicle.

Project 19: Adding High-Level Mirror-Mounted Indicators

You Will Need

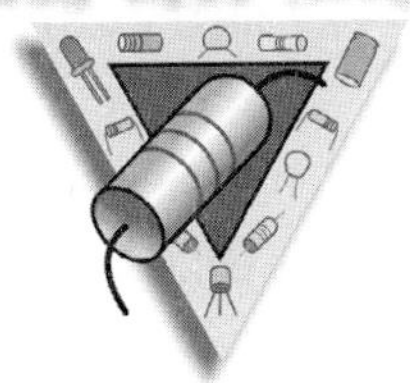

- 12 V orange LEDs
- Bezels to suit OR pre-modified aftermarket mirrors
- Door contactor plates (optional)

Tools

- Drill
- Assortment of drill bits

Many newer vehicles are fitted with high-level mirror-mounted turn signals as the safety benefits brought by increased visibility are proven.

It is possible to modify your existing mirror housings and incorporate these higher visibility signals. The quality of work produced will depend largely on your skill and manual dexterity in manipulating the plastic housings. However, if you do not feel up to the task, there are many aftermarket mirror sets available for fitting onto existing vehicles.

Many of the mirrors are supplied with a generic base plate, which fixes to the vehicle. Adapters for specific vehicles are also available. However, if one is not available for your exact vehicle, it is probably easier to modify or fabricate a bracket for the mirror than it is to modify your existing mirror housing.

Two types of high-level turn signal are available. The forward-facing variety incorporates a clear strip on the front side of the mirror housing. Although you may have the skill to construct an indicator strip in the front of your mirror housing, it would generally be said that it is easier to buy an off-the-shelf item. The forward-facing high-level mirror-mounted indicator is epitomized in the M3 style of mirror that is readily available, popular, and seen on many modified cars. The mirror is shown in Figure 4-1.

Another approach to the forward-facing turn signal would be to drill a series of holes in which LEDs in grommets can be inserted. With suitable filing and blending it should be possible to make the LEDs look fairly unobtrusive.

The other type of turn signal is a rear-facing one that is integrated into the face of the actual mirror. When not in use, several metallic blips can be seen on the surface of the mirror. When in use (that is, when indicating before turning), an arrow is illuminated in your mirror making your intentions clear to vehicles behind you. This is illustrated in Figure 4-2.

If your mirrors are plastic, you will find them relatively easy to drill. If, however, your

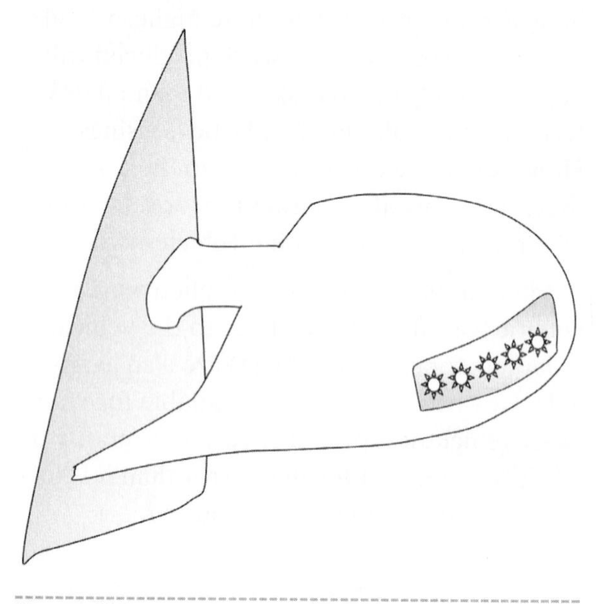

Figure 4-1 *Front-viewing high-level indicators*

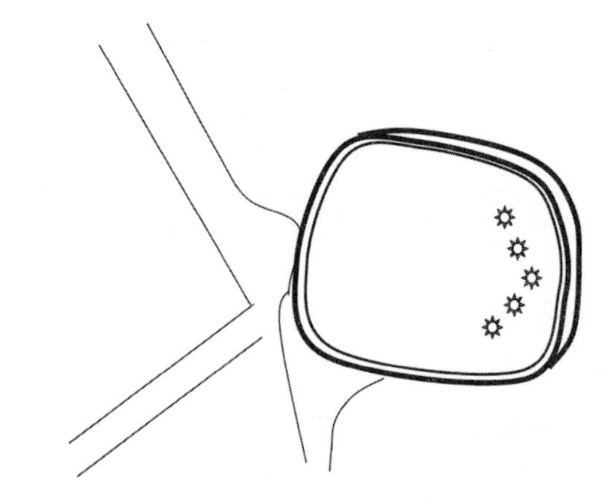

Figure 4-2 *Rear-facing high-level indicators*

mirrors are glass, do not attempt to drill with an ordinary drill bit. Use a special glass bit or seek help from a local glass retailer. The holes you drill will need to be large enough to accommodate the LEDs you choose as well as a small grommet.

When selecting LEDs, choosing a 12-volt variety will take a lot of pain out of the wiring. Although a vehicle's power is generally slightly higher than 12 volts, these LEDs seem to deal quite well with the overload. If you use generic LEDs remember, you will need a resistor in series to step down the voltage. The wiring is simply a parallel circuit. Be sure to connect the LEDs the correct way, as they are polarity sensitive.

The LEDs should have tails of flexible cable soldered on to them, with the joint protected by a small amount of heatshrink tubing.

Be sure that the LEDs have plenty of cable within the mirror housing so the mirror is able to move through its normal operating sequence. When positioning the cable, be sure that it will not snag against any of the mirror's internal workings, such as electric motors.

As the wires pass out of the mirror housing and into the door, be sure that they are free from any obstruction that may be caused by other moving parts, such as electric windows or central locking mechanisms. The cable should be routed through the door to the door-shut panel.

There are two methods of making a connection.

As the indicators do not need to be used when the door is open, a pair of plates can be purchased. One has spring-loaded contacts and is mounted to the door. The other has flat metal plate contacts and mounts to the car body. When the door is closed the plates contact each other and make a connection.

The other alternative is to run the cable through the door while it is shut and allow a little extra cable for free movement. Be sure that grommets are used wherever a metal bulkhead is penetrated to ensure that the wires do not chafe.

The connections can be made to the nearest accessible point of the indicator circuit. Quite often the flasher unit is mounted behind the dashboard. This or the indicator switch might be a suitable point for connection. Failing this, the wires can be continued and a connection made under the fender to the side repeater lights.

Troubleshooting

Troubleshooting

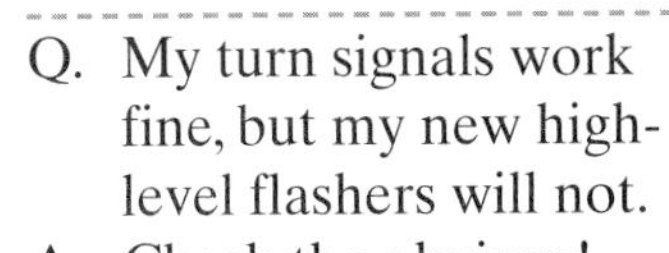

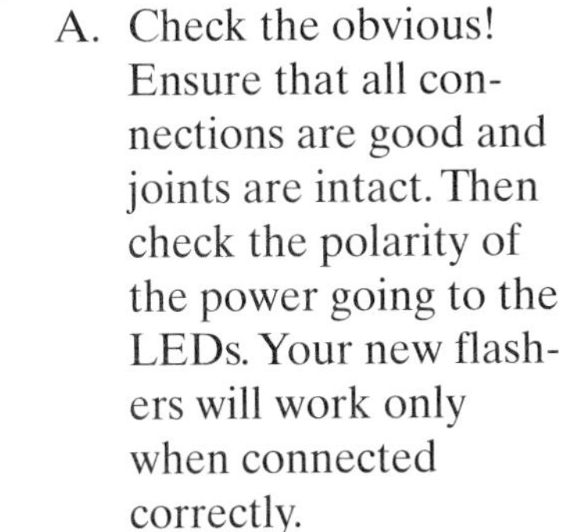

Q. My turn signals work fine, but my new high-level flashers will not.

A. Check the obvious! Ensure that all connections are good and joints are intact. Then check the polarity of the power going to the LEDs. Your new flashers will work only when connected correctly.

Q. My turn signals now flash too fast.

A. You are probably using an older mechanical flasher whose flash rate depends on the load. Switch to a newer solid-state flasher whose flash rate is determined electronically. This should cure the problem.

Project 20: Adding Puddle Lights

How many times have you stepped out of your car in the dark only to find a puddle of mud or some other unpleasantness awaiting the approach of your shiny new shoes?

Onerous conditions and the lack of streetlights often prevent you from seeing things at night. Puddle lights will help to illuminate your path as you leave the vehicle.

Puddle lights are becoming more common on some high-end European vehicles. When you open the door, a small light pointing toward the ground from the bottom of your wing mirror illuminates for half a minute or so. This allows you to see where you are about to walk.

The circuitry is pretty simple and consists of a group of LEDs and a time delay provided by an RC network.

You Will Need

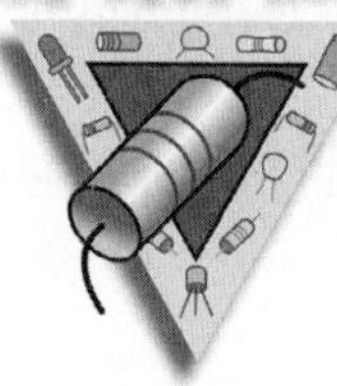

- Ultrabright 12 V white LEDs
- Grommets, sized to suit LEDs
- 1K resistor
- 47K resistor
- Current-limiting resistor, for LEDs to suit
- 470 μF capacitor
- 1N 4001 diode (×2)
- BC 548 transistor
- 12V relay

Tools

- Soldering iron
- Side cutters
- Drill bit, to suit LEDs and grommets

The circuit is assembled according to Figure 4-3.

When installing your LEDs into the mirror housing, be sure that their installation does not prevent any of the mirror's mechanisms from working. You will find that the LED's leads will need to be flattened against the mirror housing if there is limited room inside the enclosure. As it is rare that LEDs fail, there is no harm in attaching the leads permanently to the mirror housing using some form of epoxy resin or potting compound. The wires from the LEDs should be run inside the vehicle with the other mirror wires, ensuring that they are not exposed to any sharp edges of the vehicle's bodywork. The circuitry can be mounted in a small enclosure, close to the courtesy lamp door open switch.

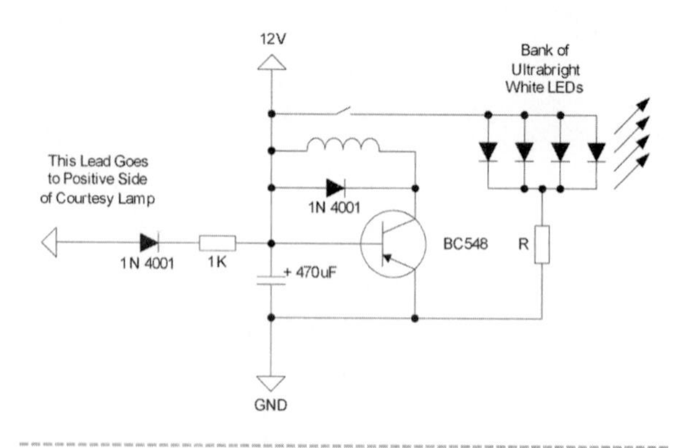

Figure 4-3 *Schematic for a puddle-light delay*

Project 21: Installing a Total Closure Module

An open window is an invitation to the opportunist thief. With increasing amounts of valuables being kept in the car and the increasing cost of car sound and navigation systems, it makes good sense to ensure your vehicle is as secure as it can be.

A total closure module can be linked to your car alarm to automatically close all of your windows and sunroof (if fitted) ensuring that your vehicle is safe and free from intrusion. This connection is relatively simple.

If you need further information about how electric windows operate, see Figure 2-6 in Chapter 2 for a wiring diagram.

The total closure module intercepts the wiring of electric window circuits thereby providing power to close the windows when a signal is received from the alarm. The total closure module includes a timer to apply power to the windows for a set period ensuring that they are closed fully.

Choose a suitable site for the module. Near to the alarm module is sensible. The module requires one connection to the alarm: the auxiliary output trigger, which should be negative when activated. The module requires power so a permanent 12-volt feed and good ground should be provided.

A variety of different modules can be found on the market. They drive a different number of motors; depending on your requirements, you will need to purchase a two-door or four-door kit and an optional sunroof kit. The wiring for all is essentially the same.

You may want to review Chapter 2. Remember that the switches are daisy chained so that both connections to the motor are held at a 12-volt potential. Because they are both at the same potential, no current can flow. The switches also have a ground feed. When the switch is depressed either way, the ground is connected to one of the motor legs. This means than current can flow and a circuit is made.

You need to ascertain what motor contact is grounded when the switch is depressed in the up position. This can be found using a multimeter.

Figure 4-4 illustrates how the window module is connected.

The wire that is grounded when the window is being wound up must be intercepted and cut, and the window module must be connected in serial.

The window module must be connected as close to the motor as possible and after any intermediate switches, child lock switches, and so on.

Keep in mind that connecting a sunroof module is essentially the same as connecting to a fifth window. The switch used to open and close the sunroof is the same as those used for electric windows, and the wiring is the same.

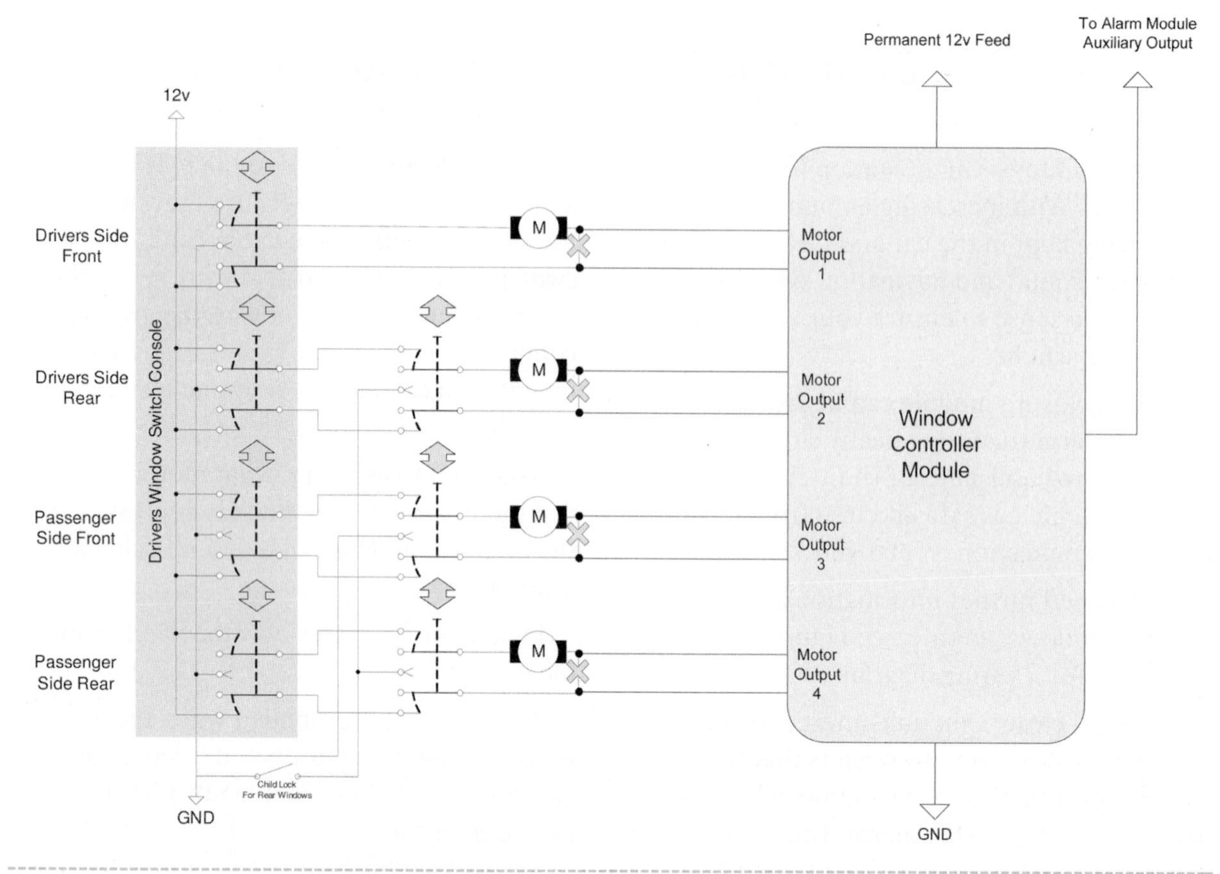

Figure 4-4 *Total window closure connection details*

Project 22: Installing Central Locking

Remembering to lock your doors is not only a nuisance; it's also avoidable! By installing central locking, you can negate the need to individually lock your doors.

Aftermarket central locking motors are available much like the one shown in Figure 4-5. These systems comprise a motor, a reduction gearbox, and a rack and pinion to convert the rotary motion into a linear movement.

There are two types of central-locking motors. The master type also has an integral switch to sense what position the door is in.

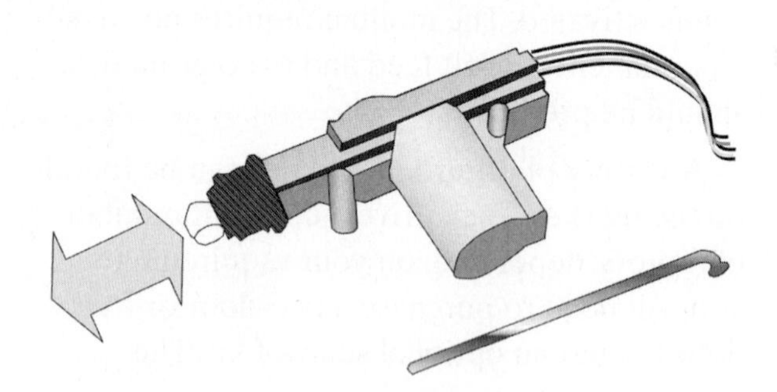

Figure 4-5 *Central-locking motor and actuator rod*

You Will Need

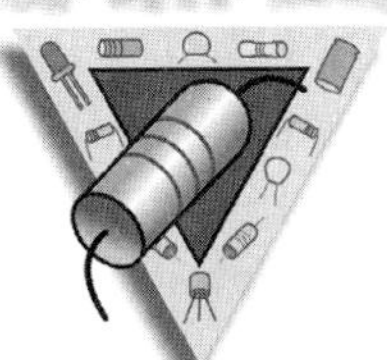

- Master-type central-locking motor for doors that have a key
- Slave-type central-locking motor for doors that do not have a key
- Push rod (for each door)
- Punched metal strip
- Fixing hardware

Tools

- Screwdrivers
- Hacksaw
- Assortment of crimps
- Crimping tool
- Side cutters
- Electric drill
- Assortment of drill bits
- Assortment of Allen wrenches

Installing Central-Locking Motors

Figure 4-6 shows as a good guide to how central-locking motors are installed.

First of all, you will need to remove the vehicle trim panel to allow you access to the lock mechanism. A great many different types of lock mechanisms exist. In this project, I am trying to be as generic as possible. You will need to identify a rod that slides when the door is shut and the door lock is activated. The door lock will be activated a number of different ways from inside the vehicle.

In many vehicles, there is a button that pops up out of the door cap when the lock is activated. The pushrod from this button runs to

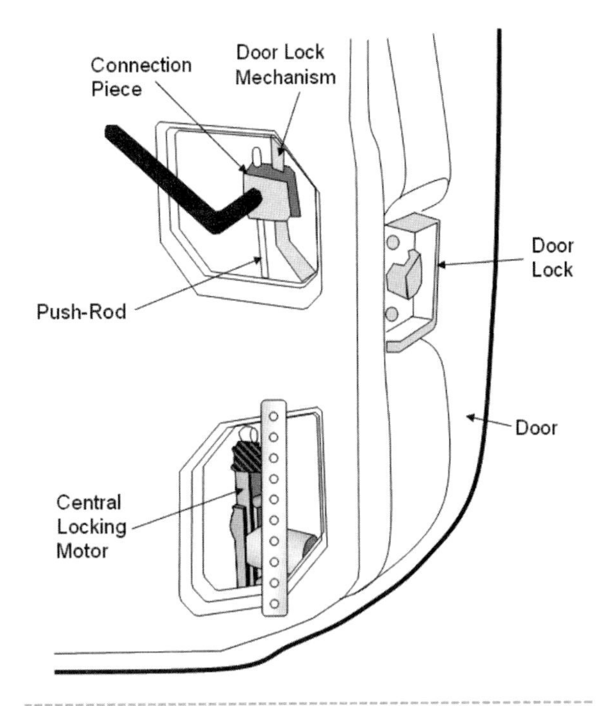

Figure 4-6 *Central-locking installation diagram*

the door lock mechanism. You will need to mount the central-locking motor, so that its action runs parallel with the aforementioned pushrod.

A variety of hardware should come with your central-locking motor. First of all, you will find a punched metal strip, not dissimilar to something from a Meccano/Erector set. This strip is used to secure the motor to the vehicle's bodywork. You should try to identify an existing hole or recess in the bodywork internal skin where the motor can be mounted.

In the event that this is not possible, you will need to cut out a hole that is larger than the bodywork in the inside door skin. Be sure to hammer over the edges so that any sharp metal faces inside the door do not pose a risk to any wiring or moving parts or to you while you are working.

Before finally securing the motor, you will need to thread the metal rod, which comes supplied in the motor hardware package. The rod will be cut at one end and have a bend in the other with a mushroom head on the end of the metal. Thread the cut end through the plastic hole and the end of the motor. Then twist the rod around so that the bent end nestles in the hole in the end of the motor. The mushroom head will prevent the rod from becoming loose. The rod should then align inside the door cavity so that it runs parallel with the door locking pushrod. It should then be trimmed so that it is the correct length.

A joining bracket can now be selected from the hardware package that came with the motor and then secured using an Allen wrench.

The motor can then be affixed with the steel plate. Again, refer to Figure 4-6 for details of how the central-locking motor should be installed.

Now we can move on to the electrical installation.

If you follow the schematic (see Figure 4-7), very little can go wrong. But once again, the important things to note are that cables should be routed correctly through metal bulkheads ensuring that they do not chafe against any sharp metallic parts, which would cause them to short out. All cables should be secured with ties or P clips to prevent them from working loose.

How It Works

When a key is inserted and turned in any of the doors with a master motor, a switch inside the motor changes position. The master unit then sends a signal to all of the other units, which move in unison, locking or unlocking the doors. The mechanical rods secure the motors to the existing lock mechanism, which is retained.

The wiring diagram for the central-locking system is illustrated in Figure 4-7.

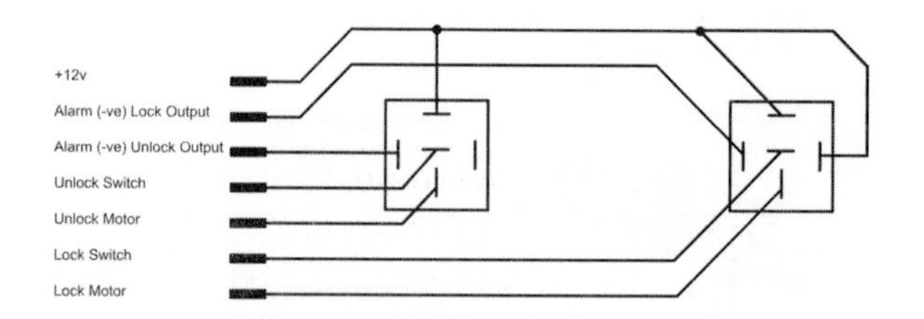

Controller Circuit is for 5 Wire 12v Alternating Positive Door Locks.
Where 3 Wire Slave Units are connected, they should be connected to "Lock Motor" and "Unlock Motor" with no connection to "Lock Switch" and "Unlock Switch".
Doors equipped with Slave Units, will not be able to control other doors in the car
And should be fitted where no external locks are present.

Figure 4-7 *Electrical installation of central-locking motors*

Project 23: Installing Heated Washer Jets

You Will Need

- Hot Shot installation kit
- Connection wire

Tools

- Electric drill
- Selection of drill bits
- Screwdrivers
- Side cutters

The Hot Shot device used in this project is illustrated in Figure 4-8. Installing heated washer jets will allow the driver to remove

Figure 4-8 *Hot Shot screen cleaning device*

snow and ice, defog the windshield, and unblock blocked washer nozzles in the winter, as well as help to remove dead bugs and road grime in the summer. Heated water is more effective than water at ambient temperature at removing detritus. Of course, heated washer jets are no good on their own—they require the whole washer system to be well maintained and in good working order.

In this project we will be covering the installation of the Hot Shot aftermarket kit by Microheat. This installation has been written for installation to a windshield washer system, although nothing should stop you from installing it on rear-window washers or headlamp washers.

Installation

Figures 4-9 and 4-10 illustrate the hydraulic and electrical installations respectively. The unit must be affixed to a sturdy mounting place; the car's bulkhead is ideal. Imagine you are looking at the unit from the front, which is to say, with the power connection pointing downward and the hydraulic connections pointing to the right.

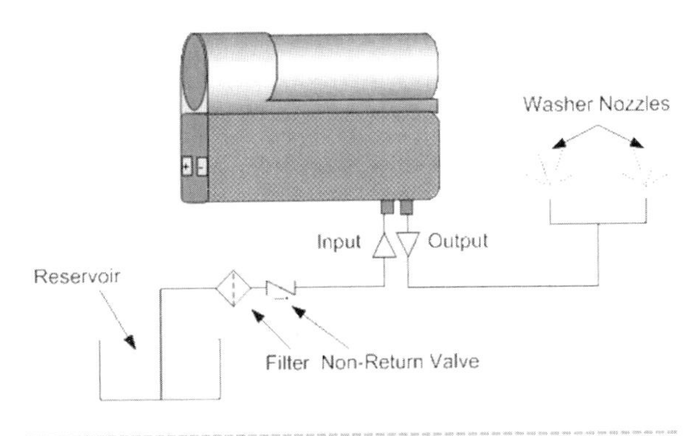

Figure 4-9 *Schematic of hydraulic connections to Hot Shot*

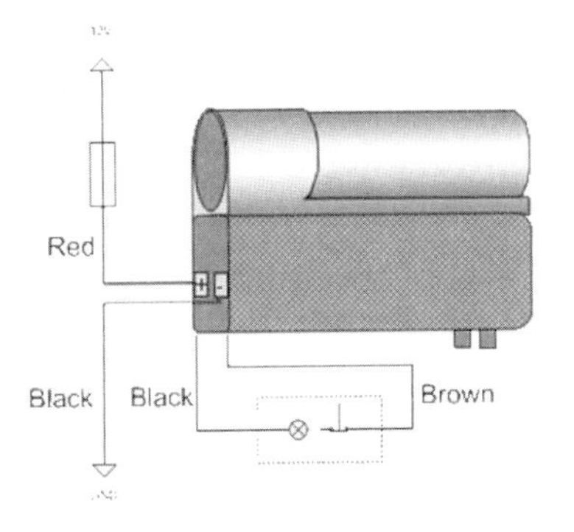

Figure 4-10 *Schematic of electrical connections to Hot Shot*

The unit must be mounted vertically with no more than 30 degrees deviation front to back, no more than 45 degrees deviation to the left (away from the hydraulic connections), and no more than 30 degrees deviation to the right (toward hydraulic connections).

Try to mount the Hot Shot as close as possible to the washer nozzles to ensure that the jet stream reaching the windshield is as hot as possible.

Do not mount the Hot Shot near any hot parts of the engine as damage may occur to the device.

The Hot Shot comes with a clip-on bracket to facilitate easy mounting to the vehicle. Before commencing installation, it is always a good idea to check that the bracket mates and to note the orientation. When you have done this, choose a suitable place and, using a fixing method of your choice (screws are recommended, bubble gum less so), secure the bracket to the bulkhead. It is now a simple matter of pressing the unit home.

Hydraulic Installation

Now for the plumbing! Find where the washer hose that supplies water to the jets runs near the Hot Shot. Cut the hose in the section that runs between the reservoir and the T junction, just before the hose separates to the two different wipers. The reservoir will probably have an existing pump in built. Except for this, no alterations are necessary. After you cut the pipe, you first need to connect the small filter supplied in the kit. A small section of supplied hose then fits on the other end of the filter. Then connect the one-way valve, being sure that the arrow is in the correct orientation to allow fluid flow. The valve can then be connected with the input from the Hot Shot unit and the output to the washer nozzles.

CAUTION

The input and output ports are supplied with yellow caps covering them. Remove these prior to installation!

Electrical Installation

The supplied power cables are rather chunky to handle the current draw of the heating element. Secure them to the unit with the supplied ¼-inch eyelet and route them sensibly to the fuse box or other suitable connection point. The red power cable comes supplied with an inline fuse holder; make sure that this is somewhere that is easily accessible. Be sure the fuse is fitted.

You now need to install the control switch in a suitable location and route the black and brown cables to the unit. A connector is supplied to allow you to route the cables through the firewall. The connector needs to be assembled. To do this the terminals should be inserted and pushed home until you hear a click. The strain relief is then snapped shut.

Warning

Before the unit can be used, it must be primed with water so it is ready and to avoid overheating the heating element. To do this, turn your key to on and spray water through the system as you would do normally until water squirts out of the washer nozzles.

Operation

The microheat module has two modes, automatic and manual.

Automatic Mode

While in the automatic mode, the module heats the fluid for just under a minute and provides half a dozen or so applications of hot fluid to the screen. The device will then shut down after a couple of minutes or if told to by the driver.

To activate this mode, the switch must be depressed for a couple of seconds until the light starts to blink. This mode is ideal in severe weather conditions, or if trying to remove ice and snow when parked. It will also help to defog the screen.

Standby Mode

If you press the button for less than a second, the light will come on and stay on. This indicates that the Hot Shot is in standby mode, heats the fluid to a predetermined level, and maintains it at that temperature within the unit. When the washer button is depressed, the fluid is heated again. Standby mode is used to remove the buildup of grime and foreign matter on the blade. It will also remove snow and ice.

Project 24: Building a Homing Device

You Will Need

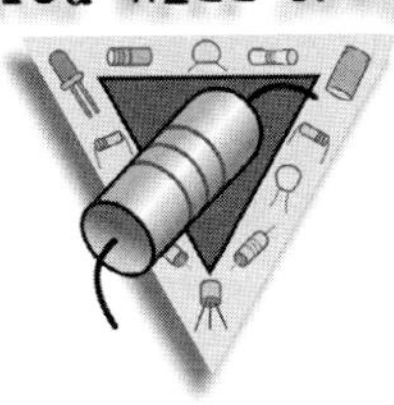

- 555 timer IC
- 741 op-amp IC
- LEDs (×7)
- 33R resistor
- 47R resistor
- 100R resistor
- 150R resistor
- 220R resistor
- 390R resistor
- 470R resistor
- 1K resistor (×2)
- 22K preset resistor
- 1M preset potentiometer
- 0.1 μF capacitor
- 47 pF capacitor (×2)
- 10-100 pF capacitor (×2)
- 500-1,500 μH inductor (×2)
- Length of wire for aerial

Tools

- Soldering iron

Picture the Scene

You parked your car at the supermarket, but where? You remember that it is a red car, but then lots of red cars are in the lot! Suddenly the parking lot begins to go all blurry as you realize you have forgotten where your car is parked.

Thankfully help is at hand! This homing device uses James-Bond-style technology to allow you to locate your car with ease, and the beauty of it is . . . it's incredibly simple to construct.

Figures 4-11 and 4-12 show the transmitter and receiver for the homing device.

The circuit operates as follows: The inductor and capacitor operate as a tuned circuit, which means that they will allow only certain frequencies to pass. Therefore extraneous signals are filtered out, and only the signal from your car's homing transmitter will be detected. This filtered signal then passes on to a signal strength meter, which monitors, you guessed it, the strength of the signal. Thus as you get closer to the car, the signal strength increases. The inductors and capacitors on the transmitting and receiving units are matched so they are receptive to the same frequencies.

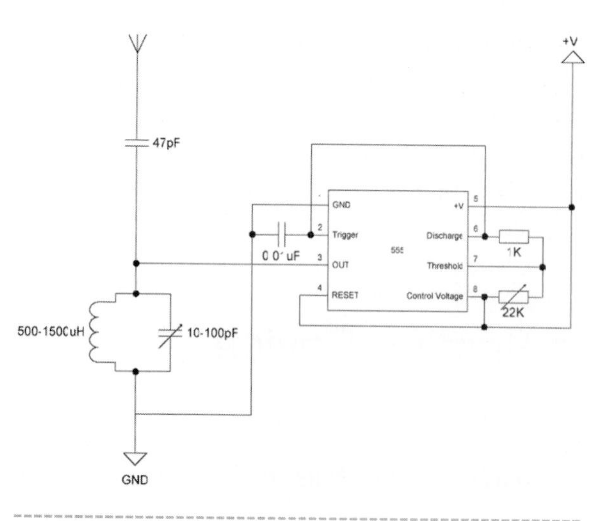

Figure 4-11 *Homing device transmitter*

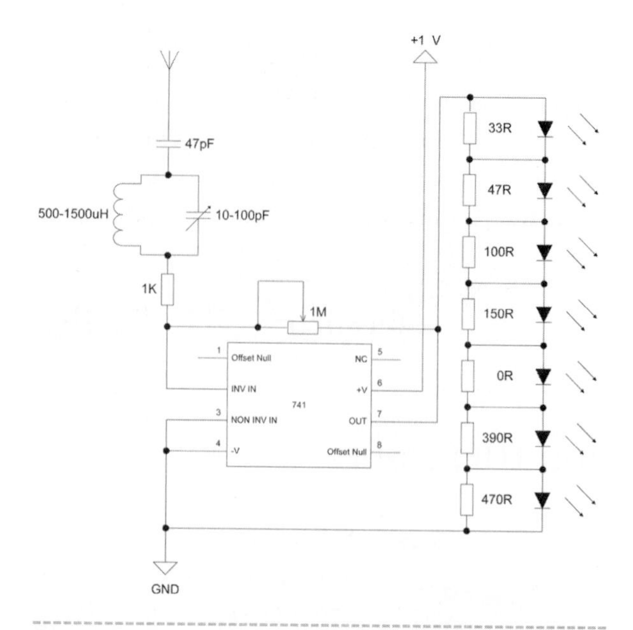

Figure 4-12 *Homing device receiver*

Warning

Please note: Check with your local regulations to ensure that operation of this circuit is legal in your locality. Certain frequencies are reserved. These change depending on local laws. Failure to do so may result in you breaking the law. The author assumes no responsibility for unlawful/irresponsible use of this device. It is possible to change the frequency on which this device operates by making a substitution of the inductor and capacitor in the transmitter and receiver.

Project 25: Installing a GPS Vehicle Tracker

It is a sad fact, but vehicles continue to be stolen every day; many disappear without a trace and with little chance of recovery. Car alarms and immobilizers are a good deterrent to the thief, but once they are disabled, little can stop the vehicle being driven away. Installing a blackbox tracker allows you to discover where your vehicle has been driven once it has been recovered. Additionally, this is useful when trying to catch the thief, as their movements and whereabouts can be determined by the blackbox.

The Deluo blackbox tracker is known as an off-line tracker, as it does not constantly transmit signals. Instead information is downloaded when the vehicle is recovered. Online trackers, on the other hand, are expensive and require a monthly fee.

The device comes with a Bluetooth interface, so it is not even necessary to gain access to the vehicle to retrieve the tracking information. The device is pictured in Figure 4-13.

You Will Need

You Will Need

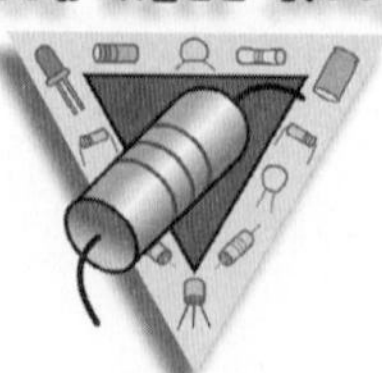

- GPS blackbox tracker
- Bluetooth dongle (optional)
- Screws (optional)
- Double-sided tape (optional)
- Velcro (optional)

Figure 4-13 *Vehicle tracking device (Courtesy Deluo)*

Tools

Screwdriver

Mechanical Installation

Mechanical installation of the blackbox GPS is thankfully relatively simple. The unit comes with two screw holes, which you can use to secure the unit to a bulkhead. If this is not desirable, it is possible to use double-sided tape or Velcro to achieve a similar end, although it will make the device easier to remove. The tracker unit can be installed anywhere, but it is highly recommended that it is placed somewhere unobtrusive where a potential thief would not think to look. As with any GPS device, the external antenna must have a clear view of the sky in order to send a signal. The dashboard and rear parcel shelf both provide suitable locations.

Electrical Installation

The electrical installation of the hardware is incredibly simple. There are two wires coming from the device, one red, one white. The red wire has an inline fuse so it must be run so the fuse is accessible should it need to be replaced. The red wire goes to a permanent live feed. It must not be disconnected when the vehicle ignition is switched off. The white wire goes to a suitable grounding point.

Software Installation

Software installation is relatively easy. Run the application bbm_setup.exe from the supplied installation CD. You will be presented with an installation wizard. Click Next, enter a location in the box, click Next again, and then click Finish. Installation of the software is now complete. The installation procedure is pictured in Figure 4-14.

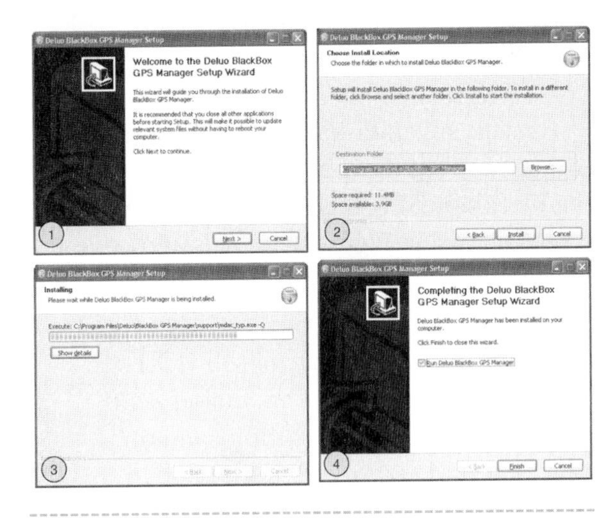

Figure 4-14 *Software installation procedure for Deluo blackbox GPS (Courtesy Deluo)*

Software Configuration

To start the Blackbox GPS Manager, go to Start > Deluo Blackbox GPS Manager > Deluo Blackbox GPS Manager. This will bring up the main screen.

The software allows you to generate reports of various functions (shown in Figure 4-15) pertaining to vehicle activity. Especially useful is the Maximum Speed function, which allows you to monitor the maximum speed the vehicle attained. This is useful if you intend to lend the car to the kids! Of more interest to commercial users is the Stop Time function, which allows you to probe how long your employees stopped during the course of their day's work—useful if you need to crack the whip on a fleet of travelling salesmen!

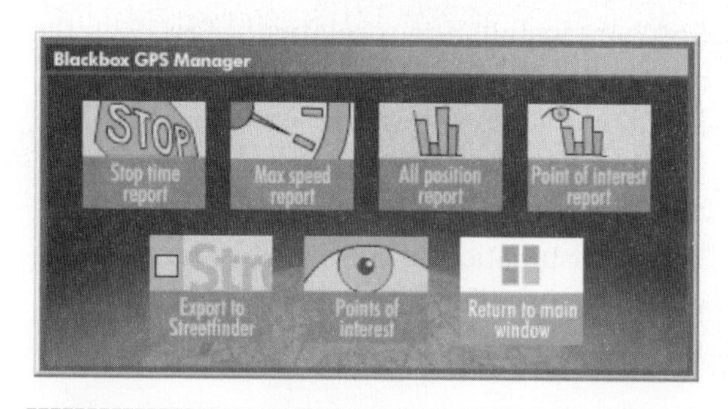

Figure 4-15 *Blackbox reports (Courtesy Deluo)*

As with any data-logging application, you will need to set the sample rate. In the case of the blackbox GPS, this can be done using the screen shown in Figure 4-16. This allows you to set how often the vehicle position is checked.

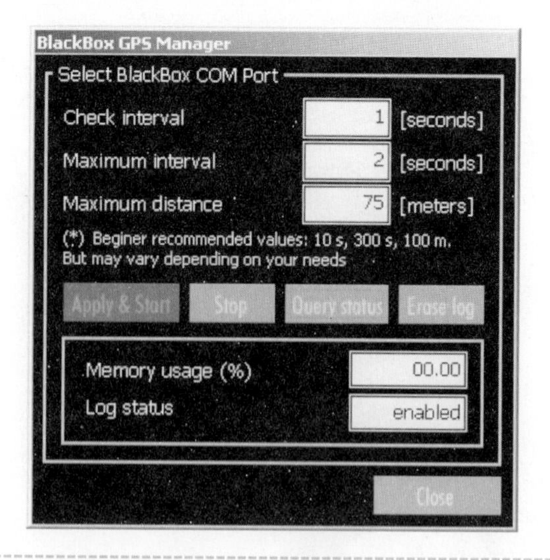

Figure 4-16 *Managing the blackbox settings (Courtesy Deluo)*

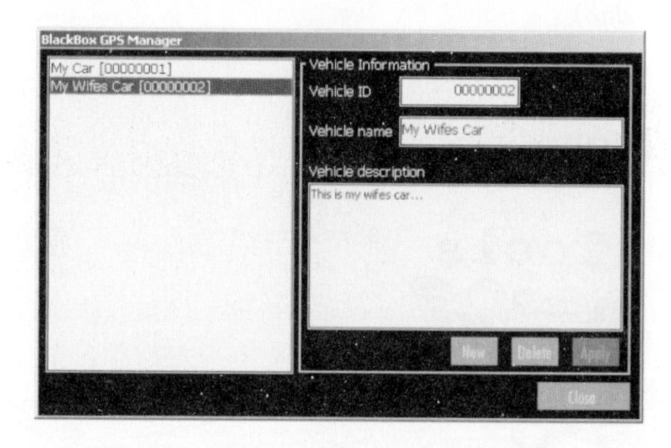

Figure 4-17 *Managing the vehicles tracked by the blackbox GPS (Courtesy Deluo)*

Clicking on Vehicle Administration brings up the screen in Figure 4-17. If you are administering multiple vehicles, for example, you may have all of the family cars or commercial vehicles fitted with a tracker. Each vehicle is given its own unique user ID number.

Project 26: Installing a Reversing Camera

You Will Need

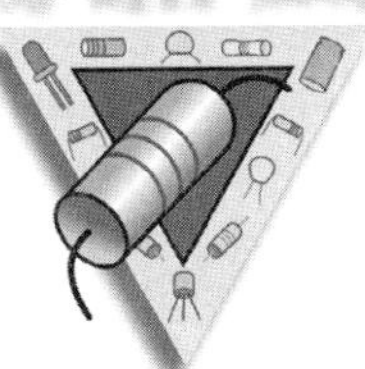

You Will Need Rostra 3.5-inch monitor (flip down) with minicam OR Rostra 2.5-inch monitor (rearview-mirror-mounted) with minicam

Tools

Tools

Screwdrivers

Electric drill

Selection of drill bits

Hole cutter

Installing a reversing camera gives you a great deal of additional safety when towing your vehicle or reversing into tight spaces. Conventional rearview mirrors have blind spots, as the car's bodywork obstructs your view. Fortunately, rearview cameras do not suffer from this problem, as they are mounted under the bumper, giving them a clear unobstructed view.

Rostra sells a great camera that can be inconspicuously mounted underneath the bumper (see Figure 4-18).

Installation is quite simple, consisting of a number of screws or nuts and bolts. You will need to locate an existing hole or, failing that, drill a new one in the bodywork. The cables from the camera should pass into the vehicle as soon as possible. Where they penetrate the bulkhead, a rubber grommet should be used.

Installation of the monitor will differ depending on the type you have selected. In the case of the rearview-mirror-mounted monitor, you will need to remove your rearview mirror and install the monitor in its place. The monitor is shown in Figure 4-19.

In the case of the flip-down monitor, shown in Figure 4-20, you will need to affix the monitor to the inside of the roof, using a single screw secured through the mounting bracket.

Once these steps are complete, the electrical installation is simply a matter of connecting a power source and connecting the composite video feed from the camera to the monitor.

Figure 4-18 *Under-bumper camera mounting (Courtesy Rostra Precision Controls)*

Figure 4-19 *Monitor (Courtesy Rostra Precision Controls)*

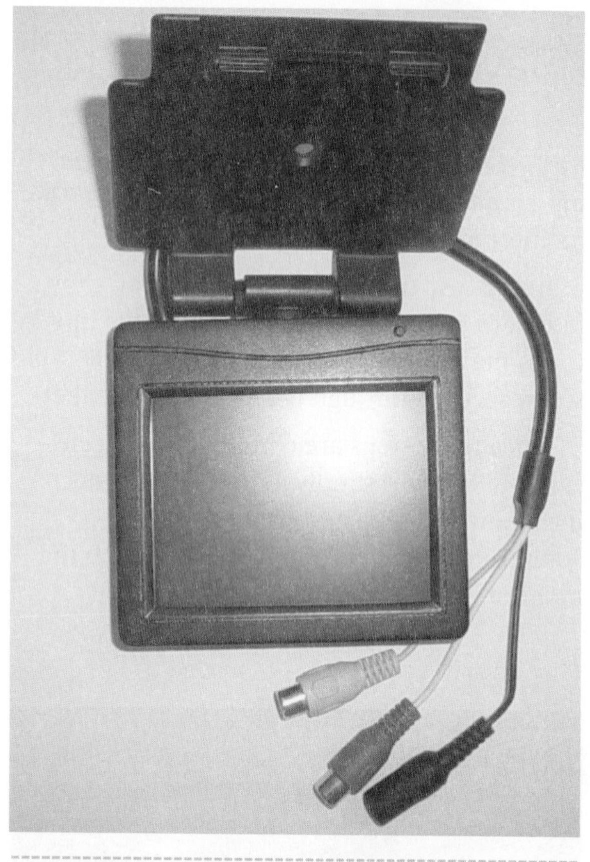

Figure 4-20 *Flip-down composite monitor (Courtesy Rostra Precision Controls)*

Project 27: Installing a Rear-Obstacle Sensing System

You Will Need

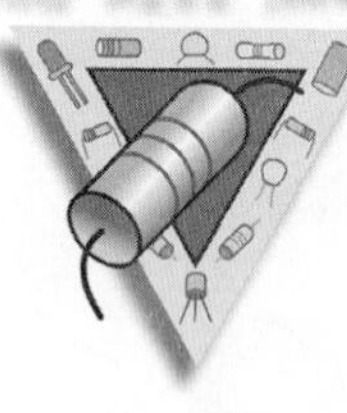

- Sensor or module assembly
- Audio control module
- LED extension harness
- LED assembly
- Hardware package
- Audio control extension harness

Tools

- Screwdrivers
- Drill
- Drill bits
- Side cutters

For this project I have used the rear-obstacle sensing system (ROSS from Rostra). For the wiring harness installation of the ROSS unit, look to Figure 4-21 for guidance.

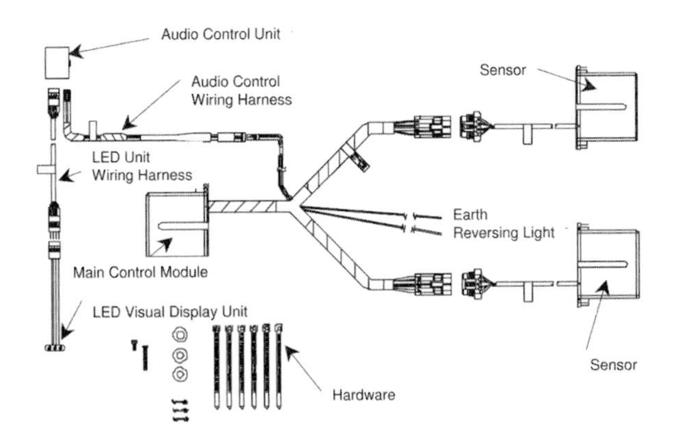

Figure 4-21 *Schematic for ROSS wiring harness installation (Courtesy Rostra Precision Controls)*

Before you start installation of ROSS, make sure that your vehicle is parked properly and that the parking brake is applied. If the vehicle is an automatic, it should be in the park position.

The sensors are mounted on the rear of the vehicle and connect to the main ROSS module. Near to the driver, there is an audio control unit and an LED display that must be positioned within sight of the driver.

You will need to mark the sensor-mounting positions on the rear of the vehicle. Ideally, the sensor should be mounted at least 14 inches from the ground. To decide where to position the sensors, measure the width of the vehicle and divide it by three to find the interval width. The sensors should be placed at two 1/3 intervals from the side of the vehicle. For example, if your vehicle is 60 inches wide, you will place the first sensor 20 inches from the side and the second sensor 40 inches from the side. When positioning sensors, some tradeoffs may need to be made depending on what materials the vehicle is made from. For example, you may find it is easier to mount a sensor in a plastic bumper than in a metal trunk lid.

It is important that the sensors are mounted parallel to the ground. The sensors should both be at the same height.

CAUTION

If a vehicle is fitted with a tow bar, hitch, or trailer hookup, the sensors should be mounted above the height of this fixture.

Warning

The sensors should be no closer than 5 inches from the exhaust pipe or any other moving items that could give false readings.

The sensors have a slot in the casing. This should be mounted so that it faces away from the ground. The sensor must be mounted so that the sensing face extends at least 1/4 inch from the bumper or any other part of the vehicle that makes contact with the sensor. However, be careful not to position the sensor too far from the vehicle, as this will leave the sensor more prone to damage.

When securing the sensor, be sure that the bolts are tight so that the sensor cannot move, rotate, or come free from its mountings. Two methods exist for sensor mounting. The first involves inserting a bolt with a washer inside the slot of the sensor housing and one outside the sensor housing. A nut tightens the washers and sensors and secures the mounting. This mounting method is illustrated in Figure 4-22.

The alternative method of mounting the sensor is to use a shorter nut and bolt with a lockwasher. This attaches to a perforated strip that can be used to secure the sensor. This method is illustrated in Figure 4-23.

Hint

The sensor is waterproof, which gives a lot of flexibility with sensor mounting.

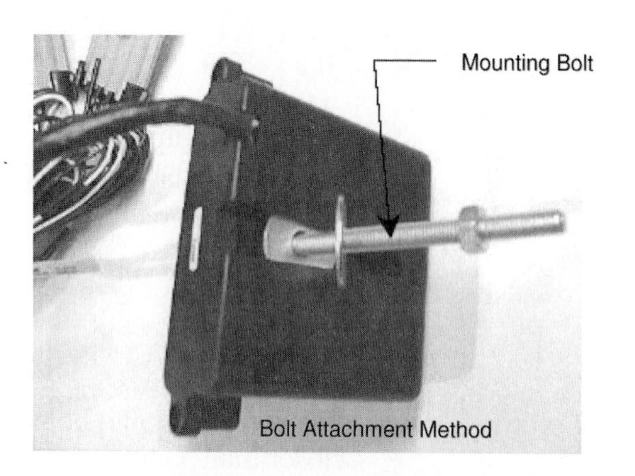

Figure 4-22 *Mounting the sensors with bolts and washers (Courtesy Rostra Precision Controls)*

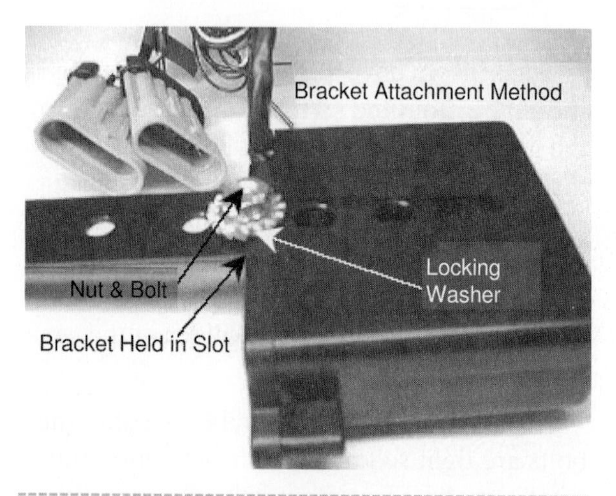

Figure 4-23 *Mounting the sensor with a bolt and lockwasher (Courtesy Rostra Precision Controls)*

You will need a ¼-inch hole in the vehicle bodywork or bumper to allow you to mount the unit securely. This is very easy due to the fact that Rostra provides the unit with color-coded connectors. The green connector connects to the green connector, and the red connector connects to the red connector. The harness should be secured to the vehicle bodywork with cable ties and P clips, or as preferred.

Warning

Ensure that the cables are kept well clear from the exhaust pipe and any other hot or moving parts.

You will need to locate the wires that lead to the backup lamps. The wire that runs from the transmission switch to the backup light(s) will need to be tapped and connected to the blue wire on the Rostra harness. The black wire will need to be connected to the other wire, which leads to ground. Or failing this, a good vehicle body ground must be provided.

The next item that needs to be installed is the audio control unit. This unit must be secured inside the vehicle in such a way that the driver is able to hear its output. It should not be mounted behind any bulkheads or padded items that would prevent the driver from hearing the audible warning. Speaker grills provide an ideal way of mounting the unit unobtrusively. This consists of connecting a 2-pin connector from the main vehicle wiring harness.

The audio unit has a small hole where a Philips screwdriver can be inserted to adjust the preset volume potentiometer. If the audio unit needs to be mounted in an awkward location where the cables cannot reach, they can be extended with some signal wire.

Once this step has been accomplished, the LED display unit needs to be mounted. This connects electrically to the audio control unit. By positioning this unit in the rear window, the driver can see it easily when looking in his or her rearview mirror when backing up.

On vehicles with a third (or rear window) brake light, the housing for the light provides an ideal mounting location, as it means that the rear windshield is not obstructed further.

An adhesive strip can be found on the LED unit and will permit it to be attached directly to the rear windshield. The windshield should first be cleaned thoroughly with some alcohol to ensure that no dirt, grease, or grime remains that would prevent incorrect adhesion.

Installation is now complete.

Operating the ROSS

For the ROSS to operate successfully a number of criteria must be met:

- The vehicle's ignition should be in the on position.
- The vehicle should be in reverse.
- The object to be detected must be moving.

As a result of the type of sensor ROSS uses, the distance at which obstacles are sensed may be influenced by the object's density and moisture content.

The ROSS's sensor coverage is divided into a number of zones. This is illustrated in Figure 4-24.

When an object intrudes upon the safe zone, the green LED will blink but no audio warning will sound.

When an object intrudes upon the hazard zone, the green LED will be on constantly, the amber LED will blink, and the audio alert will beep slowly.

As the object encroaches on the hazard zone, the green and amber LEDs will be constantly lit, the red LED will blink, and the audio will change to a fast beeping tone.

Finally, when the danger zone is crossed, all of the LEDs will be lit, and a continuous tone will come from the audio output.

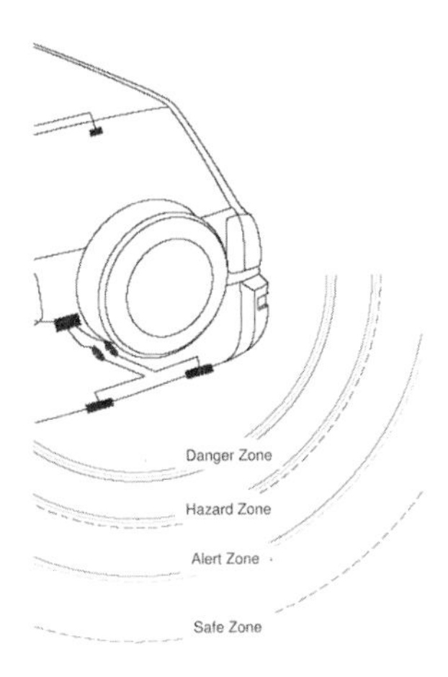

Figure 4-24 *ROSS detection rear zones (Courtesy Rostra Precision Controls)*

Project 28: Constructing Dome-Light Dimmer Delay

An interior lamp adds a sense of security when you are entering and exiting the vehicle. On many modern luxury vehicles, an interior dome-light delay allows the lamp to stay on for a few moments after you shut the doors *and* has theater-style dimming. This adds a real tangible sense of luxury to the vehicle, as the interior lights dim down as if it were the start of a premiere performance!

This circuit takes it one stage further. It provides for two-stage dimming. This allows you to have two separate interior dimming circuits (for example, footwell lights mounted under the dashboards and under the front seats as well as interior lights mounted in the roof of the vehicle). This allows the footwell lights to dim some time after the interior lamps are off. Various effects can be achieved within the vehicle.

The circuit is very simple and can be achieved with a pair of op-amps, a pair of power transistors, and a handful of discrete components. The circuit diagram is shown in Figure 4-25.

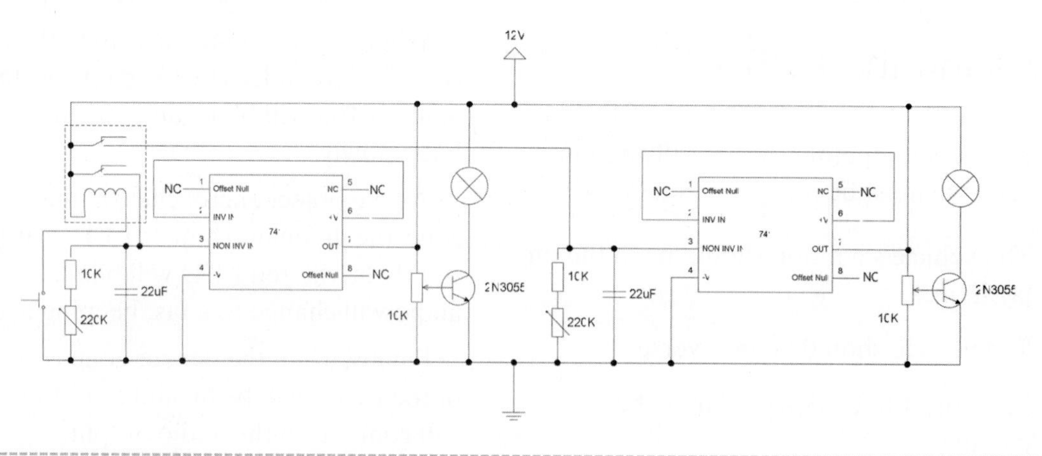

Figure 4-25 *Circuit diagram for theater-dimming lights*

Chapter Five

A Computer in Your Car

Car Computer Basics

Within the field of car telematics is a business area that has rapidly expanded in the past decade, and trends predict that it will continue to do so. In the past, advanced in-car features were only available as a factory fit option, but this chapter and the ones that follow will show you some of the things you can install and build yourself (see Figure 5-1).

In an automotive setting, we use the word telematics to describe an integrated communication and *information technologies* (IT) system that is used to enrich the passenger and driver experience in the vehicle. Because of the vast amount of flexibility offered by modern PCs, your in-car telematics system can be as simple or complex as you like.

One of the benefits of building it yourself over buying a commercially made unit (not to mention cost) is that the PC architecture is very modular, allowing you to add standard components or remove them at will. Many commercially available in-car entertainment products do not offer this flexibility.

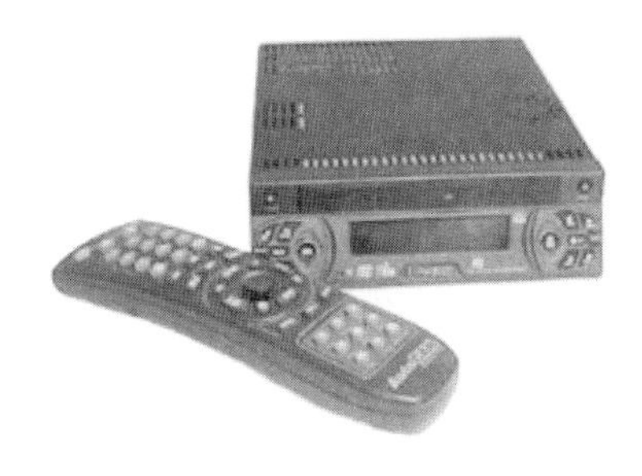

Figure 5-1 *AutoCAN car PC (Courtesy AutoCAN)*

Also, spend a few moments on the Internet and take the time to consider how much a PC DVD drive and an in-car DVD player cost. You will see that a great disparity exists between the cost of the in-car system and the PC drive, the PC drive being a fraction of the cost of the in-car DVD player. Yet when the PC DVD drive is used with a car PC, the car PC gains the same functionality as the much more expensive in-car system. Start adding more items to the list—a *global positioning system* (GPS), an in-car Mp3 player—and it quickly becomes apparent that an in-car PC makes good economic sense.

In this chapter, we will be giving a broad introduction to car PCs, what you can achieve with prebuilt solutions, and what you can construct yourself. If you are not keen on getting too involved with the "nuts and bolts" of car PC construction, a number of commercially available car PCs are on the market, and if you have the space in your car a number of "mini-PCs" could comfortably fit in the trunk. The next chapter will give you an overview of what is available commercially and some insight into what you can achieve at home.

You can find more information online about AutoCom products at www.autocomonline.com.

AutoCom ACEX C3

The AutoCom ACEX C3 car PC system (see Figure 5.2) looks like a high-end amp when you take it out of the box. It is a solid, sturdy unit and weighs only 7 pounds, so it will not add any significant weight to your car.

One of the advantages of the ACEX C3 system is that it has no need for an operating system, as some functionality is built into the read-only memory (ROM) of the VIA motherboard. This means that for some applications time-consuming boot processes are avoided, the PC being ready to use in a matter of seconds.

The AutoCom ACEX C3 is as flexible as any other car PC. Its dimensions are 2.5 × 8.25 × 10.25 inches, which should allow even the smallest automobile to accommodate its diminutive frame.

The monitor is not integrated into the case; therefore, the user has a variety of mounting

Figure 5-2 *AutoCOM ACEX car PC (Courtesy AutoCOM)*

options. These will be discussed later in Chapter 7, "Display Technologies." In this installation guide, we will be discussing installing the foldout screen supplied by AutoCom, which will fit into a 1 DIN slot. (DIN is from Deutsches Institute für Normung, which is to say the German institute for standards.) Installing the AutoCom is simplicity itself and ideal for someone who is inexperienced with PCs and automotive electronics.

Project 29: Installing the AutoCom ACEX C3

You Will Need

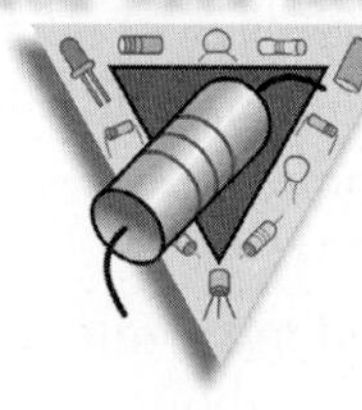

- AutoCom ACEX PC
- In-dash monitor
- Cabling supplied by AutoCom

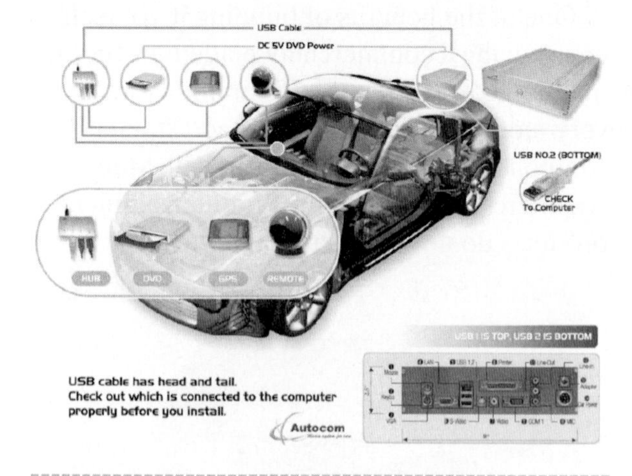

Figure 5-3 *AutoCom ACEX peripheral connection details (Courtesy of AutoCom)*

Tools

- Screwdrivers
- Side Cutters
- Drill
- Drill bits

The AutoCom ACEX C3 is mounted in an external enclosure so it does not fit into the conventional dashboard slot. The screen component of the AutoCom installation will mount in a standard DIN slot (see Figure 5-3).

To remove your old radio, you may find you need a DIN removal tool. If you don't have one, a DIN removal tool can be fabricated out of some stout metal wire. The tool is U-shaped and fits into the two holes on either side of the radio. It pushes back metal flaps that engage with the DIN surround when the radio is installed. When the prongs are in place, the radio should slide out freely. Do not force it; you may find that your installation also uses screws or another method so this step may vary a little.

When the radio has come out, be careful to disconnect the wires carefully and make sure no bare metal rattles around. The touchscreen cannot be inserted just yet. A number of connections need to be made.

A "cigarette lighter"-style 12-volt accessory socket powers the monitor. If your vehicle does not have one of these, it may be a good idea to install an inline socket using the wires that connect to your radio to power the socket. Check that the wires are of a sufficient cross-sectional area to support the current going to the PC. If they are not sufficiently thick, you may need to install a thicker feed from the battery.

Two connections need to be made to the main base unit. A *video graphics adapter* (or VGA) connection supplies the visual signal to the monitor, while a *universal serial bus* (USB) cable carries the touchscreen position information back to the base unit (see Figure 5-4).

Find a place for the base unit before deciding to route the cables. As is the mantra with any of the projects in this book, put the cable in position first of all to ensure a sufficient length before securing the unit in place. Once the monitor is secured, it is time to make the connections to the base unit.

Three connections must also be made: a positive connection to the battery, an earth connection, and an "ignition feed" connection for shutdown/startup purposes (see Figure 5-5).

The rest of the connections are to the peripheral devices. This is just like installing devices on a home PC. The hub must be connected to USB port number 2.

This is the bottom port on the ACEX C3 case.

Also, a 5-volt power connection goes from the ACEX C3 case to the external DVD drive.

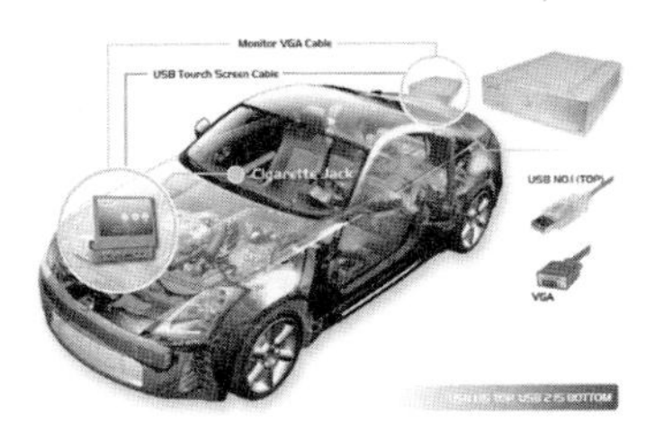

Figure 5-4 *Autocom monitor connection details*

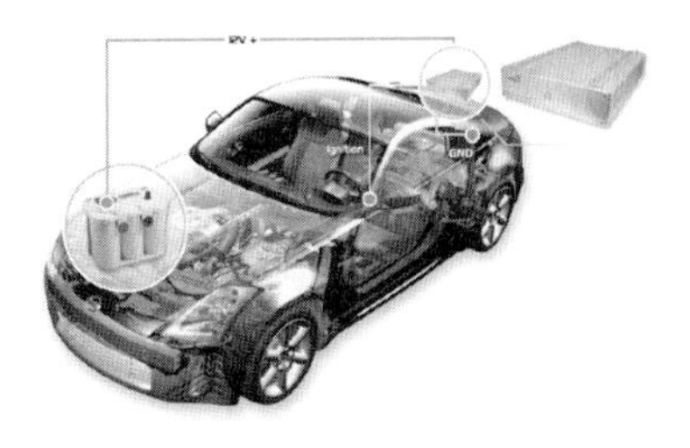

Figure 5-5 *AutoCom power connection details*

The devices that then connect to the hub are as follows:

- DVD
- GPS
- Remote

Building Your Own Car PC

If you have never opened up a PC before, now might be a good time to try! I like to think of building PCs as being child's play. When you were a child, you may have played with a shape-sorter, blocks that would only fit a certain way into a slot. PCs are very similar. A motherboard provides a wide array of connectors that are all different shapes, sizes, and colors. On the whole, there is only one way things will fit together. The golden rule is put things together gently and don't force them. If you follow this method, things are pretty foolproof. If you want a practical demonstration of this concept, try plugging headphones into a printer port. It doesn't work and vice versa.

A little knowledge is dangerous. A lot of knowledge—now that's just plain darn destructive!

Without wishing to instill you with a false sense of confidence, before attempting to build a car PC, it would be worth familiarizing yourself with the components that make up a PC. McGraw Hill publishes a number of these. Check your library.

When considering what type of computer system you are going to use for your install, the following factors must be considered:

- What materials are available
- What space is available
- What performance is required

If you already have large quantities of PC techno-junk lying around, then this may influence your decision on which components you are going to use. There is no right or wrong choice of components for a car PC; you should tailor your requirements to the performance you need and the components you have available.

Many PCs built within the last few years will have some sort of variant of the ATX standard motherboard. The Advanced Technology Extensions (better known as ATX) series of motherboards is a standard specification that lays down requirements as to the physical shape of the motherboard and the connections available. This standardization has allowed manufacturers to produce cases and motherboards that are interchangeable.

Because of their large physical size, ATX motherboards will be unsuitable for a car PC. This is not to say impossible, just not ideal. Fortunately, there is a solution: the Mini (and Nano) ITX form factors.

Mini/Nano ITX

The Mini and Nano ITX form factors developed by VIA were a revolution in motherboard thinking. Traditional motherboards were large mainly because processors were interchangeable, along with fans that required mounting. Although a large motherboard offers much flexibility and choice, in many cases they are simply too large and power hungry.

The Mini ITX motherboard redefined the standard. By integrating sound and graphics onboard, many of the PCI slots could be removed. Moreover, by soldering the *central processing unit* (CPU) to the motherboard, it was possible to do away with the large chip carriers and fans.

To illustrate the savings in size, compare the size of the processor used in a Nano ITX motherboard with a Pentium M processor used in laptops and a U.S. penny. The processor is smaller than a penny! See Figure 5-6.

By limiting processors to sensible speeds, the heat generated by the motherboard could be kept low, as well as the power consumed.

The defining features of the Mini ITX motherboard are that its dimensions are smaller than 170 by 170 mm and that it consumes less than 100 watts of power. This makes it ideal for car PC applications. For relatively undemanding applications such as playing music and GPS, a system with a speed of 1 GHz or less will be quite adequate.

The great thing about the Mini ITX form factor is that it has retained many features of the ATX standard such as connector layouts and power sockets. This makes them compatible with a wide range of PC hardware.

Processors

When VIA sat down and considered how to shrink full-size ATX motherboards onto a tiny form factor, they looked at the components that were unnecessary and took up a large amount of space. Conventional PC processors take the form of a large chip that is secured to the board using a *zero insertion force* (ZIF) socket. Although this allows the user a little flexibility with processors:

- The ZIF socket takes up a *massive* amount of space.
- Introducing another electrical connection gives room for unreliability.

To this end, VIA decided they would solder the chips directly onto the motherboard. This allows the user less flexibility over chips, but it does bring benefits.

Furthermore, VIA looked at reducing the power consumption of the processors. The Mini ITX form factor has two flavors of processors: EDEN and C3. EDEN processors (see Figure 5-7) tend to be a little slower, whereas C3s (see Figure 5-8) are relatively faster. The downside to using a C3 processor is the slightly increased power consumption. The EDEN processors use such a small amount of power that they do not even require a fan.

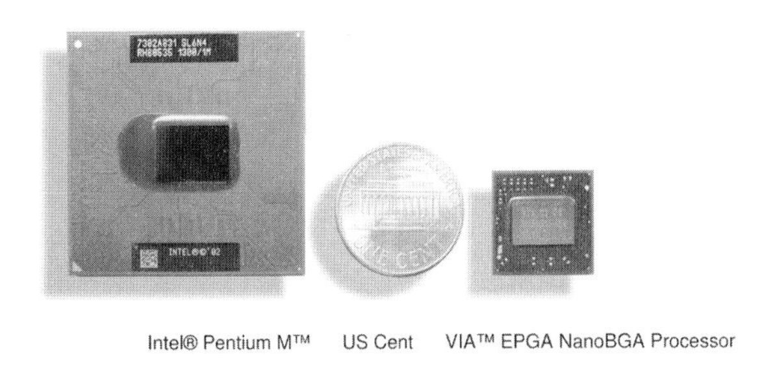

Figure 5-6 *Nano ITX processor compared to Pentium M and U.S. penny*

Figure 5-7 *VIA Eden processor*

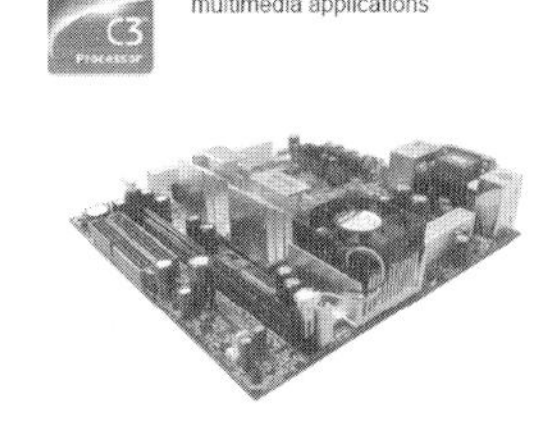

Figure 5-8 *VIA C3 processor (Courtesy mp3car.com)*

Currently, C3 processors are running at speeds of up to 1.3 GHz, with further chips in development. This speed is adequate for most car PC applications.

Memory

Mini ITX motherboards use exactly the same type of memory as a standard PC. The *single data rate/dual data rate* (SDR/DDR) is reduced depending on the motherboard that you select. With some of the more compact case styles, you will need to purchase a low-profile memory module. This shares the same connector as an ordinary memory module and functions in the same way. The only difference is it's shorter (see Figure 5-9).

Installing memory in a Mini ITX motherboard is just the same as upgrading your PC at home (see Figure 5-10). Locate the slot for the memory module, and press the tabs outward. Now take the memory module out of its static bag. Align the module so that its notches match up with the plastic moldings on the motherboard connector. Press the module firmly home, and the plastic clips should snap into place at the sides of the memory module.

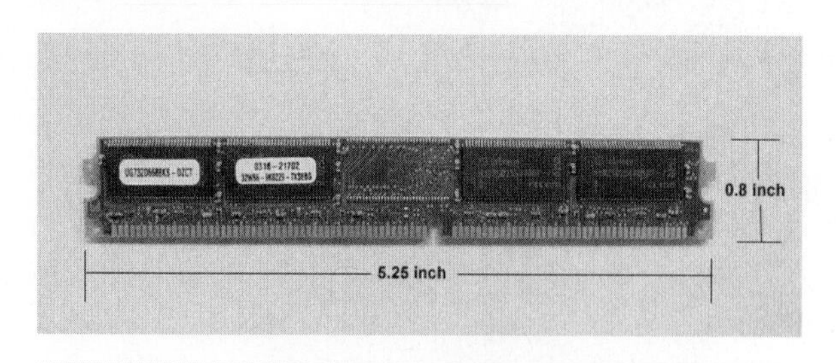

Figure 5-9 *Ultra-low profile memory module*

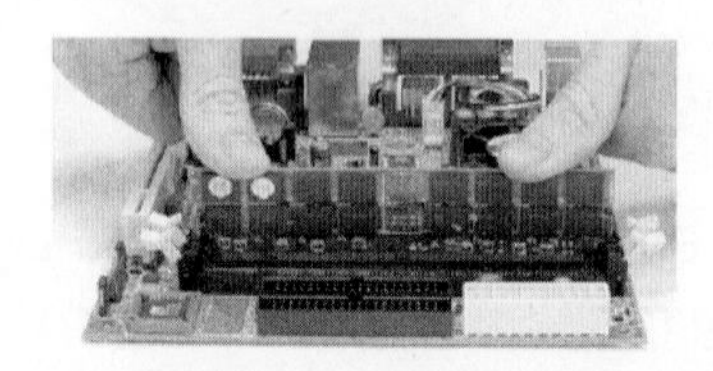

Figure 5-10 *Inserting the memory module (Courtesy VIA)*

Motherboard Options

VIA now manufactures a wide range of boards based on the Mini ITX standard, in their EPIA range. When considering the board that you want to use, you will need to think about the functionality you require, as the different boards offer subtlely different options, some that may suit you better than others. You may also find that constraints are imposed by the case that you choose, which may limit your choice of board. Have a look at the images over the next few pages (Figures 5-11 through 5-13), which clearly highlight some of the hardware features each board has to offer. The images will also serve as an invaluable reference when building your car PC, as they clearly show the locations of all major components and connectors.

Case Styles

When considering what type of case design you want and whether you are going to make one, you will need to seriously evaluate your own skills and offset the savings made in money against the time required to produce a decent end result. For the intrepid, some dimensional data is provided in this chapter that will prove an invaluable aid for those wanting to construct their own car PC case. For the less adventurous, we will have a look at some of the best cases on the market.

Once you have selected your motherboard, you need to find somewhere to house it. It is possible to squeeze a Mini ITX motherboard into a DIN radio-sized case, but for your first car PC this may be inadvisable as you want a little room to tinker and add parts as you please. A variety of cases are available on the market, and one such case is the OPUS Mini ITX case (see Figure 5-14).

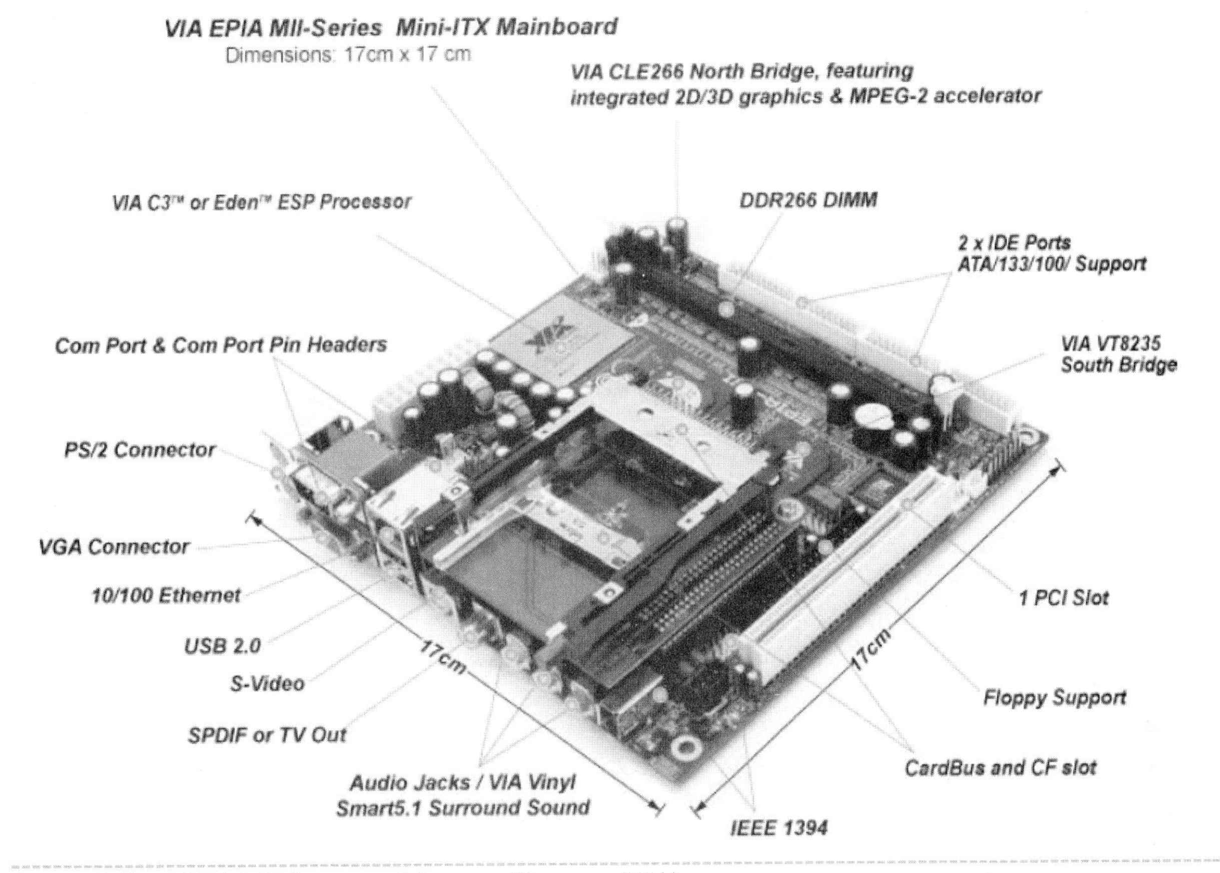

Figure 5-11 *EPIA MII layout and features (Courtesy VIA)*

Figure 5-12 *EPIA SP layout and features (Courtesy VIA)*

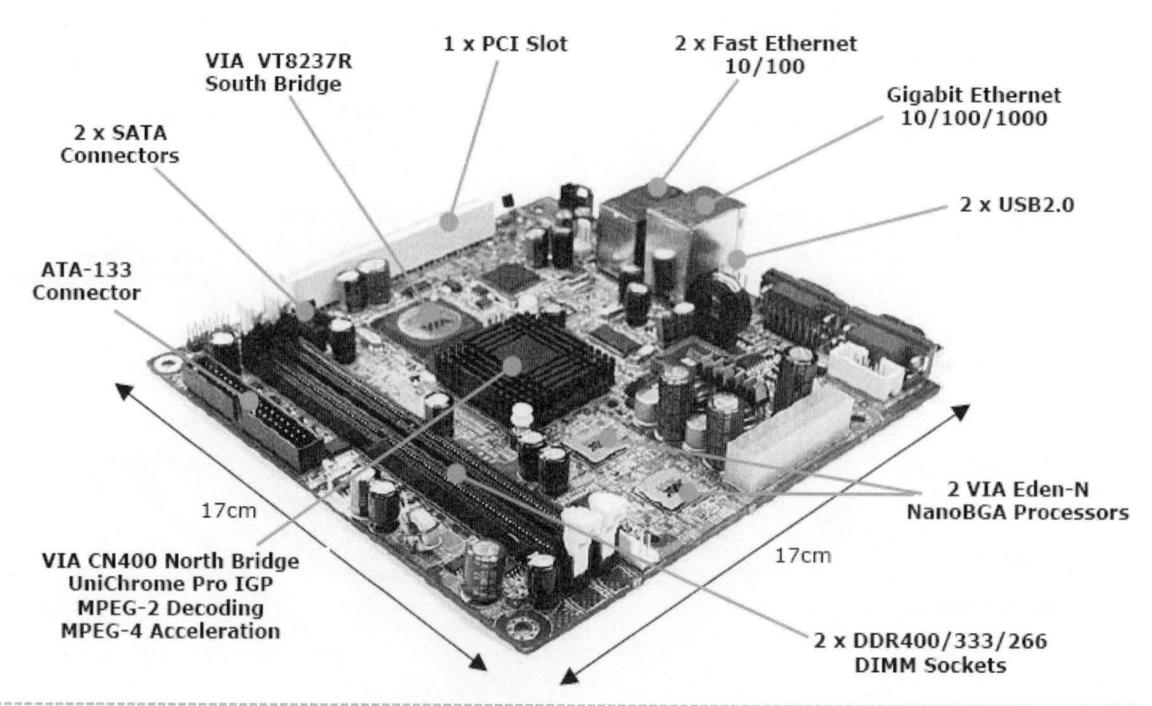

Figure 5-13 *EPIA DP dual-processor layout and features (Courtesy VIA)*

As shown in Figure 5-15, the case comes with a power supply and all fixings for constructing your first Mini ITX PC. It is available from Digital WW (listed in Appendix A). With the reputation that OPUS has built for making high-quality power supplies, the user's expectations of this case are high. Rightly so! This case is made to a very high specification and provides the valuable extra space necessary for experimentation.

Unlike many small cases on the market, the OPUS vehicle case comes with a two-PCI riser, which allows you to increase the upgradeability of your EPIA motherboard.

Figure 5-14 *OPUS Mini-ITX case (Courtesy Digital WW)*

Another option for case mounting is a standard car stereo DIN slot. The dimensions of the DIN slot are useful to know when estimating space requirements and thinking about your own car PC cases. The dimensions for a single DIN slot are illustrated in Figure 5-16. Double DIN slots are twice the height and allow the user to mount either two single DIN units or one larger double DIN case (see Figure 5-17).

For those with a little more confidence who want to accommodate their car PC in a standard DIN-sized slot, the Travla series of cases offers a very professional looking alternative. They allow you to mount your car PC in the

Figure 5-15 *OPUS Mini-ITX case (Courtesy mp3car.com)*

Figure 5-16 *Dimensions for single-height DIN aperture*

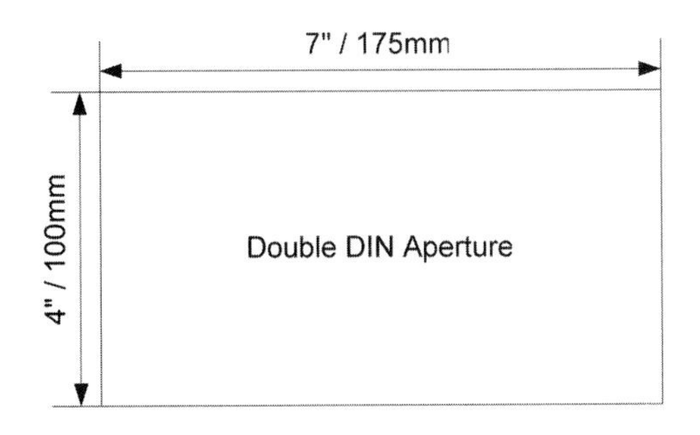

Figure 5-17 *Dimensions for double-height DIN aperture*

same size slot as your car stereo. This standard size slot is known as a DIN slot (after the Deutsches Institut für Normung).

Travla produces a variety of small cases that are able to incorporate a range of VIA Mini ITX motherboards, as shown in Figures 5-18 and 5-19. The cases are well designed and allow for laptop components, helping to keep the form factor as compact as possible.

Figure 5-18 *CI34 case, front view (Courtesy mp3car.com)*

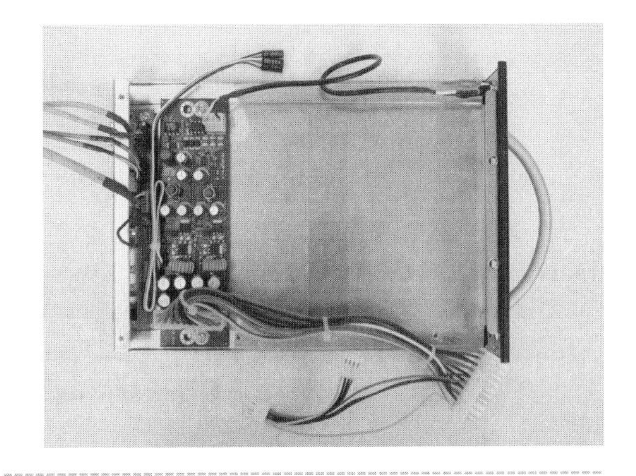

Figure 5-19 *Inside the CI34 case (Courtesy mp3car.com)*

Case Controls, Buttons, and LEDs

When you are building a regular PC with a standard case and that is supplied by ATX power, it comes with a selection of leads that connect the motherboard to the case controls. A set of case controls may be comprised of the following:

- Power switch
- Reset switch
- Power LED
- Disk LED
- Speaker (for *power-on self-test*, or POST, diagnostic beeps)

Your car PC is just the same as any other PC; it requires these buttons. Unfortunately, unless you are using an off-the-shelf case, you will need to make these yourself. Thankfully, this is a relatively simple affair.

You Will Need

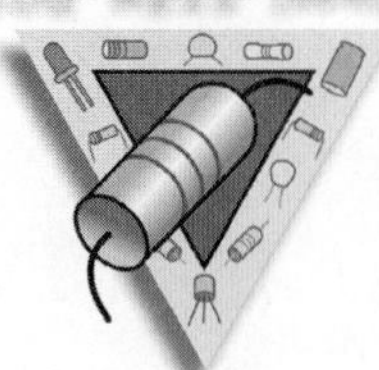

2 momentary single-pole, single-throw (SPST) switches

2 LEDs (color to suit)

Small speaker or piezo element

Hookup wire

Female D-type socket crimps

Heat shrink tubing

LED bezels (optional)

The rating of the switches is unimportant as they will be handling a small amount of power. An attractive switch can be found in the aluminum vandal-resistant switches, which have a very solid feel and nice sturdy action.

The choice of LED is up to you. It is nice if the color is similar to the dashboard illumination to maintain a consistent effect and make for a professional installation.

Some people would prefer to omit the speaker as they find the beep upon startup irritating. It should be noted, however, that it is worth making a speaker even if it is left disconnected. In the event of problems, the POST beeps will help to diagnose the fault.

The crimps are those used when assembling D-type PC connectors. Connections will vary from motherboard to motherboard. Consult with the manufacturer's information, which should clearly show the connections.

Before making the leads, decide where the buttons are to be mounted. On the dashboard? Out of sight? An appropriate position must also be chosen for the two LEDs.

Once the components have been sited, work out where the cables must be routed and how. Once the length of the cable has been ascertained, cut and solder the D-type connectors on one end with the switch or LEDs on the other. With the LEDs, observe the polarity and indicate it on the socket end of the wire.

The connections should then be insulated with heat shrink. This is less important on the switch end, but on the connector end this is imperative as headers on motherboards are closely spaced rows of pins, and an uninsulated connector could short and cause problems.

Have a look at Figure 5-20. This shows the location on the VIA motherboards of the front panel connector. This is where you should join your leads and tails.

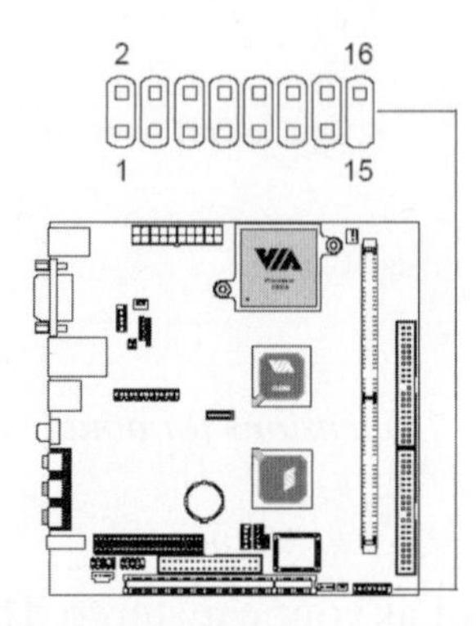

Figure 5-20 *Location of case connections VIA motherboards (Courtesy VIA)*

Troubleshooting

Q. My LEDs won't light . . .

A. Check the polarity of the connections, and try them with the leads reversed.

Constructing Your Own Case/Modifying a Standard Project Case

With the wide variety of project cases available from hobby stores, it should not be too hard to house your car PC in a custom or standardized case. In the event that your budget is very tight, a case can be improvised out of a sturdy, insulated box. My first car PC was made from an A5 magazine box file that cost £4 at the local stationers! When mounting the motherboard, use the illustrations in Figures 5-21 through 5-24 as a guide.

Drill holes and mount the motherboard securely using plastic spacers. Preferably, you should choose a case made of plastic or another nonconducting material, but if you are

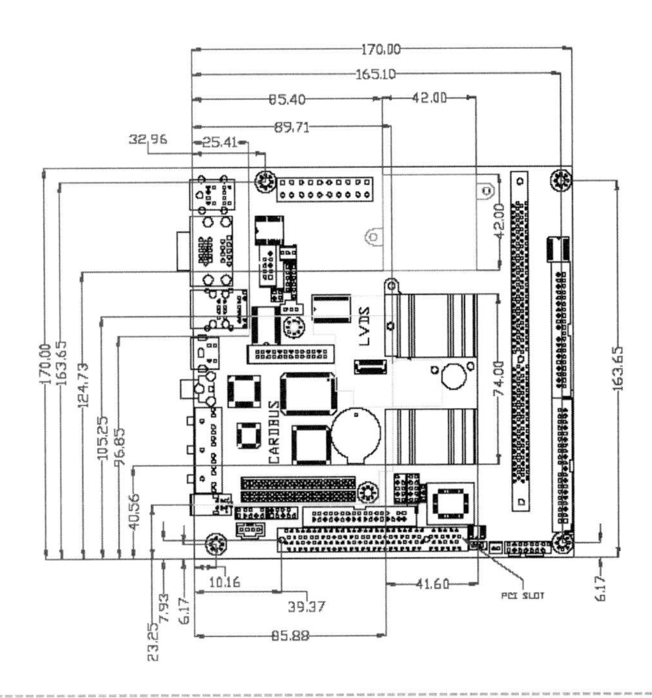

Figure 5-21 *EPIA MII motherboard details for fixing when constructing case (Courtesy VIA)*

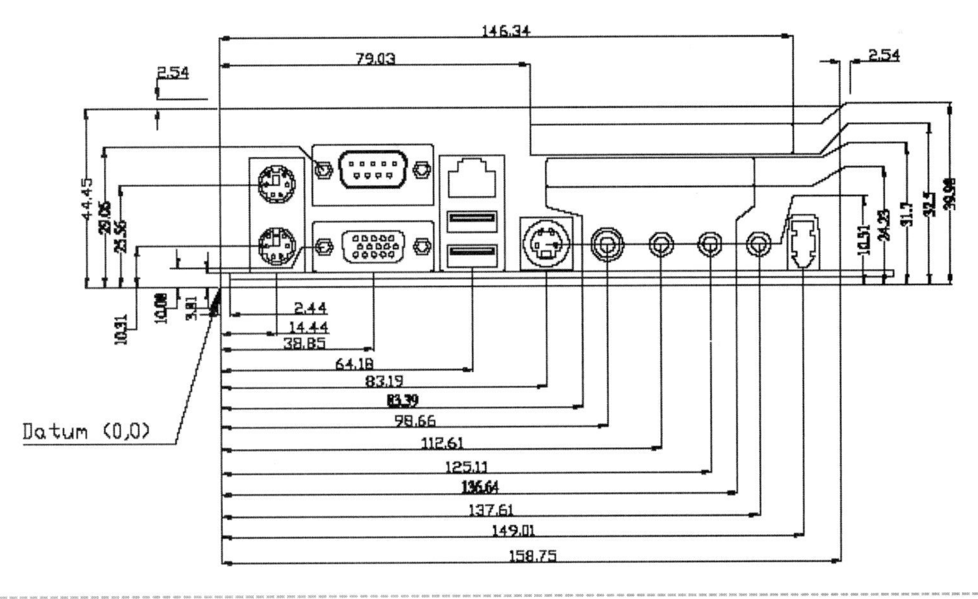

Figure 5-22 *Dimensional details of MII series motherboard connectors for layout when constructing case (Courtesy VIA)*

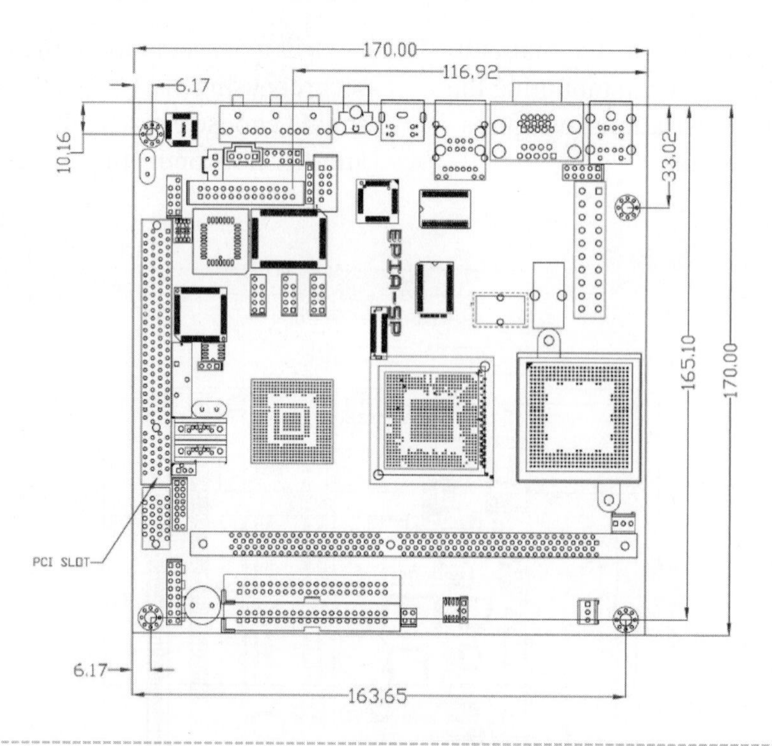

Figure 5-23 *Dimensional details of VIA SP (Courtesy VIA)*

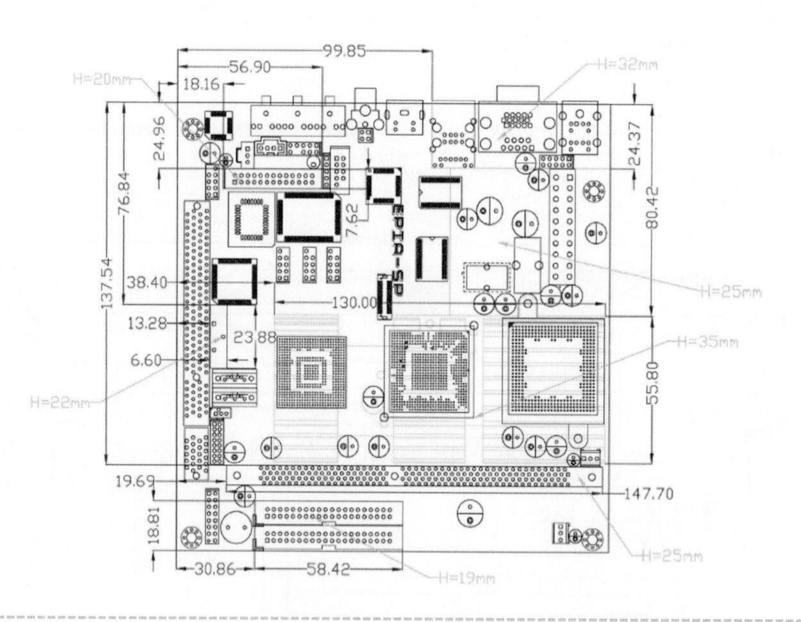

Figure 5-24 *Height and component distribution details of VIA SP (Courtesy VIA)*

going to use metal, ensure that no room exists for short circuits.

You will also need to think about air circulation and how heat will dissipate from the case. Vents and air outlets should be built in, especially if you are going to use a heat-insulating material such as wood. Think carefully about the layout of drives and what the user sees from the front of the case.

Motorbike PCs?

You may be the proud owner of a motorbike or a bubble car. If you want to build a tiny vehicle PC, then look no further than the Nano ITX form factor (see Figure 5-25). Mentioned earlier in this chapter, an application can be found in the epilogue of this book.

Figure 5-26 *EPIA-N for tiny vehicle PCs (Courtesy VIA)*

Other Car Computers for Consideration

For the die-hard Apple enthusiast, the small "Mini Mac" is one potential way of getting a Mac in your motor. As no dedicated car-power supplies are available for the Mini Mac at the time of purchase, you will need to install an inverter and plug the Mac into this. Although this is a less efficient solution, it is the only viable option.

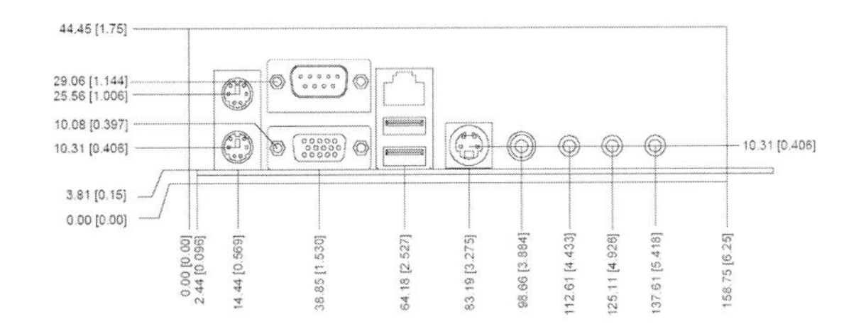

Figure 5-25 *Connector details for VIA SP (Courtesy VIA)*

Chapter Six

Power Supplies

Powering a PC inside a car is a very different affair from powering a PC at home; the supply available in a car is completely different from the domestic 230-volt AC supply. Another consideration is that the car also requires a certain amount of power for basic functions, so the full power available from the car will be reduced as other accessories are used. To help us understand the problem, let's look at the power available in a car and the requirements of a car PC.

In a car, power is stored in a large lead-acid battery. The alternator charges this battery while the car is running. When starting the car, a large current is required by the battery to turn the starter motor. In operation, the battery is required to produce much less current as the alternator does much of the work—powering the ignition system and the car's electrical systems.

The alternators fitted to modern vehicles are designed to provide power for the vehicle's essential functions and a little more for nonessential functions. Electric windows, central locking, and other additional electrical features all require power, but for a relatively short amount time. As it is not anticipated that you will be using all of the car's functions at the same, the allowance made for these electrical devices is relatively small. Unfortunately a car PC draws a sustained current as it requires power *all* of the time it is operating. In addition, if you have a high-powered audio system, you will need to consider upgrading your charging system to deal with the extra demand.

As the alternator takes over much of the work when the car is started, the amount of power in the battery never really drops to a near-empty situation, which must be avoided at all costs. There are practical considerations as well as issues of damaging the battery's chemistry if you let the battery run down. With a dead battery, the car won't start, and this could be a major inconvenience.

If the car PC is to be used on a regular basis, it may be advisable to install a split-charging system.

The amount of current a car PC draws will depend on the level of complexity, type, and number of components used. I recommend the use of Mini-ITX motherboards from VIA, as these motherboards pack a considerable punch while still drawing a relatively low amount of current.

Split-Charging Systems

Installing a split charger entails having a second, auxiliary battery fitted to the vehicle. This battery is used to power systems that are nonessential to the operation of the vehicle, such as car computers and audio systems.

When the car engine is running, the second battery is charged. When the engine stops running, the second battery is disconnected. The car PC will draw current from this battery until it is flat; this will not affect the main vehicle battery and therefore allows the vehicle to start and function normally. As the auxiliary battery is a deep-cycle battery (as opposed to a cyclical car battery), there are no problems with discharging it completely.

Project 30: Installing a Split Charger

You Will Need

You Will Need

- Leisure battery (deep cycle)
- 30A cable
- Standard 30A automotive relay
- Selection of crimp connectors
- Battery terminals
- Fusebox

Tools

- Crimping tool
- Side cutters

Assuming that the relay and car wiring follow standard DIN conventions, a standard SPST relay has four terminals labeled 30, 85, 86, and 87 (see Figure 6-1). To use such a relay as a split charge controller, it needs to be wired as follows.

Figure 6-1 *Split charger schematic*

The terminal labeled 30 needs to be connected to the positive terminal of the vehicle battery using a heavy 30-amp cable and suitable terminations. Try to keep wires as short as possible. Terminal 30 is the common connector for the relay's switch.

The terminal labeled 85 needs to be connected to ground. This is one of the connections to the relay's coil.

The terminal labeled 87 (which faces a different direction than the other three) needs to be connected to the positive terminal of the auxiliary battery.

If there is a terminal 86a, it should be left disconnected.

The ground of the auxiliary battery is connected to vehicle ground or to a piece of metal bodywork (where the bodywork is used as ground).

The final wire, 86, should be connected to the wire that runs between the alternator and the charge warning lamp on the dashboard. On many alternators, this connection is labeled IND on the alternator (presumably to represent "indicator"). If the car follows the DIN conventions, this wire will be a brown one with a black tracer.

How the Circuit Functions

When the ignition is on but the engine isn't running, the charge warning light glows, as one side is at 12-volts potential and the other side is at ground as current flows through the coil that is not energized to ground. As we have connected to the negative side of the charge warning light, the relay is not energized and no current can flow between the main battery and the auxiliary.

When the engine is turned, the coils in the alternator are excited, and the negative side of the charge warning lamp is at 12 volts of potential. The lamp is extinguished. As one side of the relay is now at 12-volts potential and the other is at ground, current can flow between the main and auxiliary batteries. This allows the auxiliary battery to charge.

When the engine is turned off, the rotor coil is not excited. Therefore the charge warning light illuminates again, and the relay contacts open, thereby isolating the auxiliary from the main battery.

Inverters

If you are looking for an off-the-shelf solution that is very easy to install, an inverter may be the answer. Inverters take the 12-volt DC supply from your battery and step it up to 230 volts. They also create a 50 Hz pseudo-sine-wave, giving 230 volts of AC that closely resembles the power available from a domestic supply.

To power a car PC using an inverter, the inverter is connected to the car's battery, and a regular ATX power supply is plugged into the inverter, just as if it was a wall socket.

Although they represent a simple solution, inverters are a messy way of delivering power to your car computer. Converting the 12-volt DC supply produced by your battery to 230 volts of AC and then back down to low-voltage DC wastes a lot of energy—most of which ends up as heat.

If you are prototyping your car PC, an inverter may be a good solution, as it allows you to ignore the complexities of a power supply while you work on your design. Although for longer term use, more elegant solutions are available.

When considering what size inverter to buy, realize that most electrical devices draw a surge current when starting up. Therefore it would be advisable to oversize the inverter to allow for this.

The inverter can be wired directly to the battery. However, this means that the inverter will be drawing current all the time. A better proposition is to wire a relay between the inverter and battery. The accessory feed from the ignition switch energizes this relay. This means that the inverter will be powered when the key is turned.

DC-to-DC Power Supplies

Unfortunately as DC-to-DC converters are a niche market, their price is high compared to an *inverter–power supply unit* (inverter/PSU) setup. Furthermore, power outputs tend to be lower at the affordable end of the market so motherboard power consumption may be an issue.

A DC-DC converter works by taking the unregulated 12 volts of DC from the car battery and then regulating it and stepping it down to some of the lower voltages required for the motherboard and CPU.

Warning

Some of the cheaper DC-to-DC converters do not regulate the 12-volt supply to the motherboard. This is so that they can exaggerate the power output of the DC-to-DC converter. This means that a 250-watt power supply could turn out to be a 50-watt regulated (the lower voltages required) and a 200-watt unregulated supply going straight to the motherboard.

The supply from a car is not particularly clean. That is, electrical noise is generated by the ignition circuitry in particular. It is unsuitable and not recommended to feed this supply directly to a motherboard. It has the potential to cause serious damage to the circuitry.

Therefore, buy a power supply only from a reputable retailer and after detailed examination of the specifications. This should prevent evil surprises.

If a power supply is cheaper, it doesn't mean that it is a better value for the money. It simply means that it is cheaper.

Startup and Shutdown Controllers

Shutting down a PC manually can be a laborious repetitive process. In a car PC, it can be further complicated by the fact that you may have car PC software laid over the top of your Windows installation, making shutting down particularly tiresome. If you leave your vehicle and forget to turn off your car PC, it will be whirring and beeping away, attracting potential thieves and potentially running down your car battery.

The flip side of the coin is getting into your car, starting the engine, getting ready to drive away ready to plan your route with the

Figure 6-2 *Startup and shutdown controller (Courtesy mp3car.com)*

on-board GPS software, only to find that you have to wait for your car PC to startup.

If you install a startup/shutdown controller (or a power supply with an integrated controller), all of this tiresome activity can be taken care of by the turn of the ignition key. The car PC starts when you want it to start and finishes when you are finished with it.

Mp3Car.com supplies a good range of startup and shutdown controllers. One example is seen in Figure 6-2.

Project 31: Connecting an External Device to an OPUS Power Supply

Unlike other "dumb" power supplies that simply control voltage, OPUS power supplies have a built-in microcontroller that monitors and regulates the voltage supplied to the motherboard.

The OPUS power supply is highly regarded in car-PC-building circles. It is a developed design that is mature and has proven itself in the field. The manufacturer has a good reputation. The power supply is shown in Figure 6-3.

The OPUS power supply offers your expensive car PC an unparalleled amount of protection against poor power conditions. The car PC is protected against double battery situations that could arise when jump-starting a car as well as against transient currents and load dumps. The power supply also maintains a clean healthy supply to your car PC when the car is being started. It does all of the following:

- It monitors the car battery to prevent a situation arising where the car will not start or function (by switching off the car PC when the battery falls below a certain level).
- It switches the car PC on with the ignition of the car, and it powers the car PC down safely and shuts down the operating system when the ignition is switched off.
- It responds to the PC's own shut-down, standby, and hibernate modes.
- It will shut down the PC automatically after 20 minutes.
- It supports "Wake-On Interrupt" to allow the car PC to be switched on via a wireless LAN.
- It provides a wake-up output for other peripheral devices that are in use with the car PC.
- It provides feedback of supply status using an LED. This makes troubleshooting easier.

Figure 6-3 *OPUS power supply (Courtesy DigitalWW.com)*

It comes in two flavors, 12 volt and 24 volt. So you may find an application in trucks or commercial vehicles where other power supplies would fail.

You Will Need

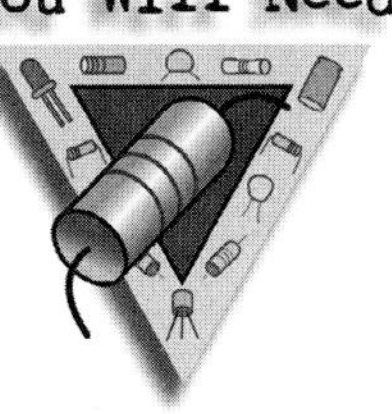

- *OPUS power supply
- *ATX power supply cable
- *Drive power cable
- *Power switch cable
- *Power supply input and ignition feed cable
- *Screws to mount

*Supplied by OPUS

Tools

Crimps

Before installation commences, you need to set a few jumpers on the OPUS power supply. This is much easier to do before the supply is fitted to the vehicle. The jumpers govern the delay between the time when you turn your ignition off and when your car PC shuts down (see Figure 6-4).

BIOS Settings

You need to set the BIOS in accordance with the manufacturer's settings. The nomenclature used for different options will vary slightly from motherboard to motherboard. Figure 6-5 shows the BIOS settings of the VIA MII motherboard BIOS settings.

The main thing you will need to do is access the *Advanced Configuration and Power Interface* (ACPI) or APM menu. Where the option is available, you will need to set the ACPI Suspend Type to S3, which suspends the system to RAM.

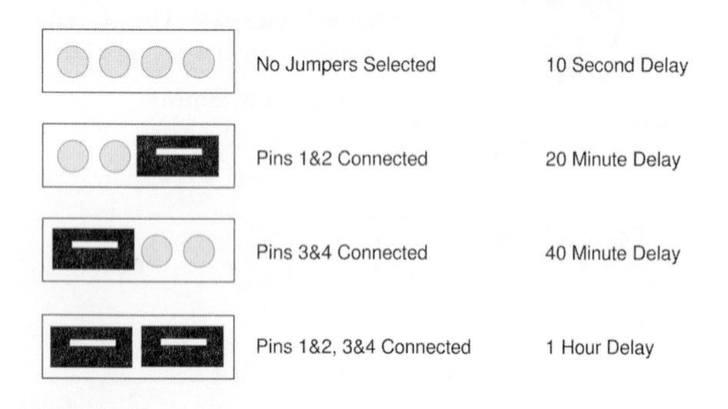

Figure 6-4 *Jumper options*

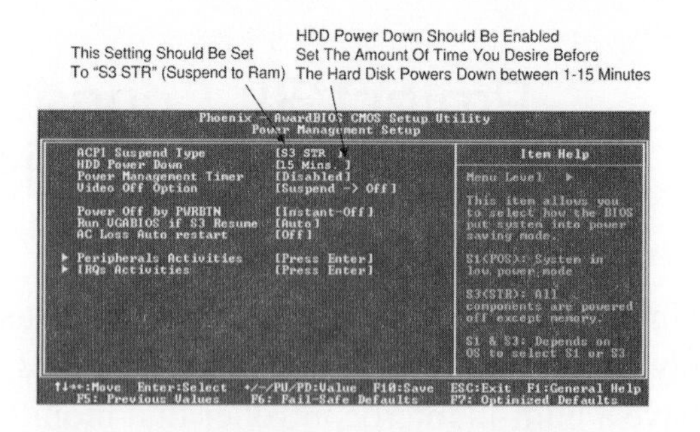

Figure 6-5 *VIA MII motherboard BIOS settings*

To conserve energy, you should also enable the HDD power-down option. With the VIA MII motherboard, this option is either disabled, or it allows you to set an integer between 1 and 15 minutes for HDD power down.

If S3 mode is not supported, the S1 mode should be used.

Where the options are available, choose the following:

- "Video repost" should be set to "enable."
- "Wake on LAN" should be set to "stay off."

Operating System Settings

You want to set the power profile of your operating system to that of a laptop or portable computer in order to reduce the power requirements of your car PC to the minimum. In Windows XP, this is done through the control panel.

Click the Start menu, and then go to Control Panel in the right-hand display column. Click on Switch to Classic View, and then click on Display. This will bring up the Display Properties box. Go to the Screensaver tab and, in the bottom of the screen, click on Power. This will bring up a screen somewhat akin to that seen in Figure 6-6.

Set your PC profile to Laptop/Portable Computer, and then, in the power buttons box,

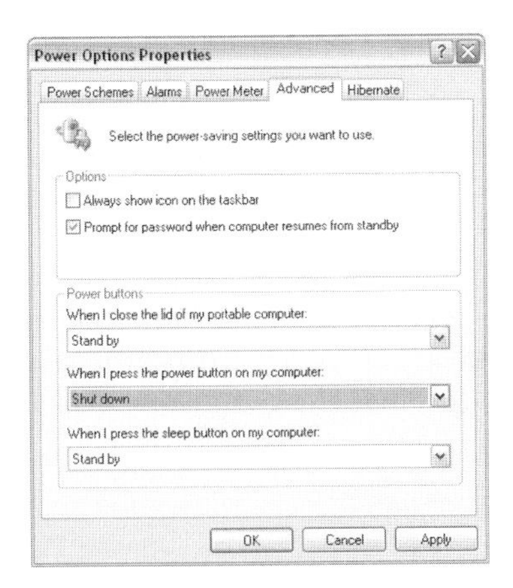

Figure 6-6 *Reducing the PC's power requirements (Courtesy Microsoft.com)*

select When I Press the Power Button on My Computer Select Shutdown. This can be seen in Figure 6-6. Then click on Apply and then on OK.

Electrical Installation

It is imperative to first locate the following wires in the car:

- A good ground—If possible tap into the car's wiring loom. If it must be attached to a panel, ensure that the joint is free from rust and a good connection is made.
- Permanent power supply feed—This is a point in the car's wiring loom that is *always* live, regardless of the position of the ignition key.
- Ignition feed—This is connected to a point on the car's wiring loom that goes live when the car is started. Optionally, it can be connected to a pushbutton switch to start the car PC "on the button."

The ATX power connector will now need to be connected to the motherboard using the supplied lead.

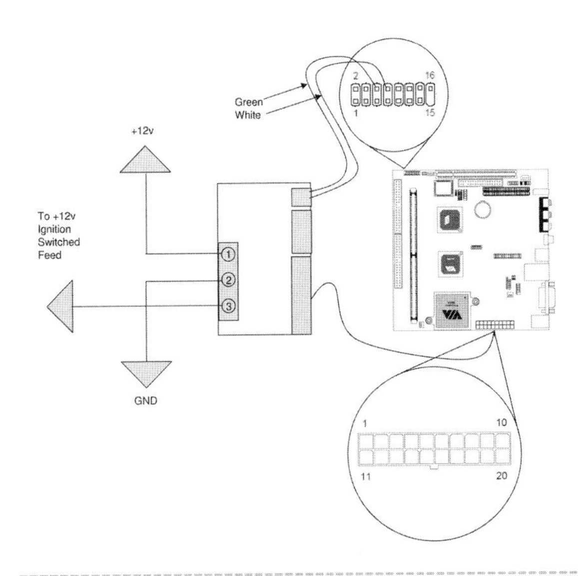

Figure 6-7 *Installation guide*

There is an additional lead; this is to be used for powering your IDE devices.

You now need to connect the power switch connectors on your motherboard to the OPUS power supply. On the VIA MII board, these are pins 1 and 2 on the front panel connector. Look at Figure 6-7 and Tables 6-1 and 6-2 for a guide to installation.

Table 6-1

Pin	*Signal*	*Pin*	*Signal*
1	PWR LED+	2	HDD LED+
3	PWR LED+	4	HDD LED-
5	PWR LED-	6	PW_BN+
7	SPEAKER+	8	PW_BN-
9	NC	10	RESET+
11	NC	12	RESET-
13	SPEAKER-	14	SLP_LED+
15	NC	16	SLP_LED-

Table 6-2

ATX power connector pinout

Pin	Signal	Pin	Signal
1	+3v	2	+3v
3	GND	4	+5v
5	GND	6	+5v
7	GND	8	PWR_GD
9	5V_SB	10	+12v
11	+3v	12	-12v
13	GND	14	PWR_ON
15	GND	16	GND
17	GND	18	NC
19	+5v	20	+5v

Operating the Power Supply

The power supply can be in a number of different modes.

When no power is being delivered, the diagnostic LED will blink briefly every three seconds or so. This indicates that the car PC is powered down and that standby current is being supplied to the motherboard. The installation will be consuming a minimal amount of current.

When the diagnostic LED shows a steady on, the car PC is on and the supply is functioning correctly.

When you switch off the OPUS power supply (by turning your ignition off), a delay elapses before the supply shuts the computer down. This supply is governed by the setting of four jumpers on the rear of the supply. The jumpers are set in accordance with Figure 6-4.

It is useful to have all devices associated with your car PC (such as monitors, drives, and additional circuits) power-up and power-down with your car PC. The OPUS will not regulate power to these devices, but it will provide a remote output to switch them on and off.

This remote output must not be used to power the devices directly, but it can be used to drive a relay that will switch the power supply to the device. OPUS manufactures a range of point-of-source power supplies to supply LCD monitors and accessory circuits. These can all be controlled from the remote output.

Troubleshooting

If you find that your car PC will not switch on or boot up, the first step is to check your wiring to ensure that power is reaching the OPUS power supply from the battery. Check that the power feed and ignition feed wires are not reversed. Also check the power supply pin headers to be sure they are connected properly to the OPUS power supply. These are the equivalent of your power-on switch, and if they are not connected nothing will work!

If, conversely, you find that the PSU will not switch off, check the diagnostic LED. Its flash patterns should give you some indication of what may be the problem. The LED will emit a few short flashes followed by a wait of 3 seconds or so.

If it emits one flash followed by a pause, the PSU is in standby mode. This could be for a number of reasons. Check your vehicle battery voltage. If it is below 11.5 volts, the PSU may have disabled your car PC to prevent further power drain.

If you find two short flashes, the PSY is indicating that the correct power-up sequence has failed. When the PC boots, it sends handshaking signals down the ATX power connector to tell the supply it is booting properly. If that has not happened, check to be sure the power supply pins are connected properly. Try reversing their polarity. Also check that you have implemented the motherboard BIOS settings correctly. If neither of these worked, you may be the victim of a locked system of a software error. Failing this, the supply can be diagnosed as faulty.

If there are three flashes, the supply is trying to tell you that its output voltages (the ones it is sending to the motherboard) are outside normal operating parameters. Check that the supply is correctly rated for the motherboard and you are not overloading it. If you have many disk drives connected, try unplugging some. Also if you have a multitude of USB devices, which drain power from the motherboard, try disconnecting some of these. If none of the above works and the supply is correctly specified, you can diagnose the supply as faulty.

Four flashes indicates that the PSU power-down sequence has failed. There can be many reasons for this. The power switch pins may be wired the wrong way with the incorrect polarity. The motherboard BIOS setting may require checking. Pay specific attention to the *advanced configuration and power interface* (ACPI) BIOS settings to ensure this function is enabled. You may also need to install ACPI drivers for your motherboard if you have not already done so. Your operating system may not automatically support this.

Also, you may find that certain software you are using is not compatible with ACPI operation. If this is the case you will not be able to use this software with the OPUS power supply.

If none of the above works, it may simply be a system hang-up or software bug. Remove power and try again. If even this doesn't work, you can diagnose your PSU as faulty.

High-quality DC-to-DC power supplies are available from OPUS at www.opussolutions.com/

Two suppliers of car PC power supplies include www.digitalww.com www.mp3car.com

ATX Power Connection

A description of the ATX power supply may prove to be useful if you need to diagnose any problems with your power supply or if you are an experienced electronics enthusiast who is intending to construct your own PSU.

For extended design information on ATX power supplies go to: www.formfactors.org/developer/specs/ATX12V%20PSDG2.01.pdf

Chapter Seven

Display Technologies

Visual display technology has come along immensely in the past few decades. A wide variety of displays are now available, and the technology is easy to implement in your own car PC.

The type of display you choose will depend upon the applications you want to run on your car computer. If you are playing MP3s and require a compact display that will integrate into your dashboard, then a simple 16×2 text display may be adequate. If you want to watch DVDs, a text display will obviously be inadequate.

When considering the size and resolution of graphical displays, it is imperative that you consider what type of applications you will want to run: It may be perfectly acceptable to watch video clips on a small 4-inch screen. But try running navigation software on a small screen, and the text becomes illegible.

For most purposes the 7-inch widescreen format has become the display of choice. It will allow the Windows desktop to be viewed with regular-size icons. It also provides ample room for the display of maps in GPS software and allows street names to be read. VGA screens in this format are now commonly available and come with the option of touch-screen (see Chapter 8 for an in-depth discussion of input devices).

Thankfully, as LCD manufacturing techniques have come of age, and the economies granted by large-scale production have come to fruition. It is now possible to buy a full-color LCD screen for less than you would expect.

To make an informed choice about which screen technology you require, this chapter will discuss the applications, limitations, and suitability of a variety of screens. First we will discuss display technologies, followed by display formats.

Display Technologies

Liquid Crystal Displays

Liquid crystal displays (LCDs) are low-cost display devices that find many applications in the automotive arena.

LCDs come in many flavors. Liquid crystal technology can be used to make simple one-line numeric displays or large full-color screens.

How LCDs Work

Polarized light will not travel through two polarizing filters that are at 90 degrees to one another. LCD displays consist of exactly that —two glass polarizing filters that are at 90 degrees to one another. Liquid crystals are sandwiched between the filters; these crystals are "twisted" in their natural state and allow light to pass through the filters. Each of the glass plates is coated with a transparent conductive material. When an electric current is applied to the two conductors, the liquid crystals' molecules align from their twisted state. This prevents light from passing between the two polarizing filters, and causes the pixel to appear black. See Figure 7-1.

Liquid crystals do not emit any light of their own. They operate on either reflected light or are backlit by a source such as LEDs or electroluminescent panels.

Vacuum Fluorescent Displays

These provide higher contrast than do LCD displays, as they are active display devices and emit their own light. They are often used in video recorders and HiFi equipment (see Figure 7-2).

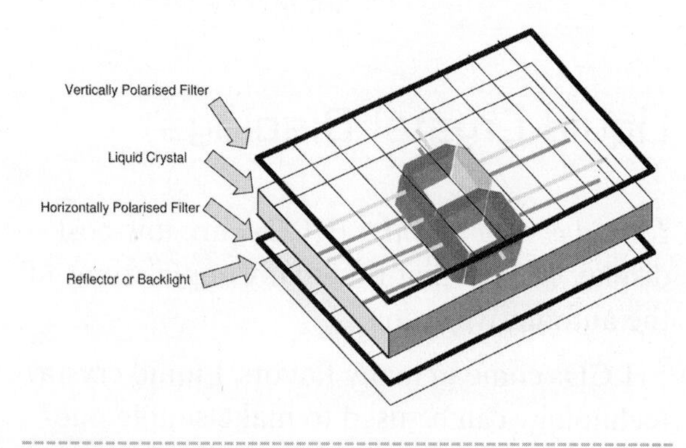

Figure 7-1 *How LCDs work*

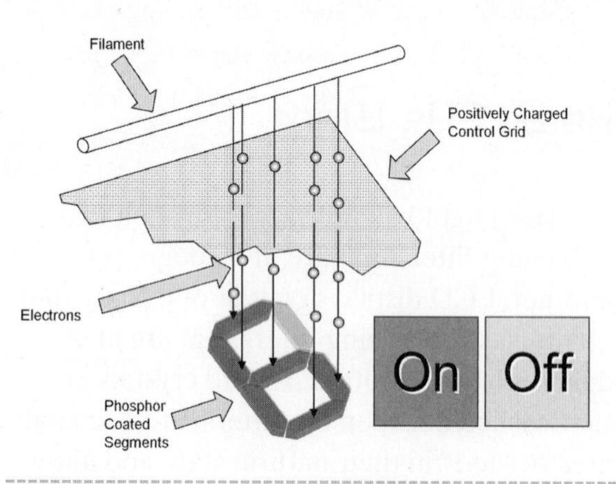

Figure 7-2 *How vacuum fluorescent displays (VFDs) work*

How VFDs Work

The dots that make up the pixels are formed from a conductive material and coated with a phosphor powder. These phosphors fluoresce when struck by electrons. A hot filament generates a cloud of electrons. When the pixel is made positive, electrons flow across the gap due to the properties of thermionic emission. Between the pixels and the filament is a control grid. By varying the voltage on the grid, the brightness of the display is changed.

Display Formats

Alphanumeric Displays

A wide range of alphanumeric displays are available. The size of an alphanumeric display is defined as the amount of characters per line multiplied by the number of lines. Thus, a 16×2 alphanumeric display is capable of displaying 16 characters per line over 2 lines.

LCD alphanumeric displays are commonly available with backlighting, which allows them to be viewed in the dark. However for some applications, a *vacuum fluorescent display* (VFD) may be preferable. Some car-PC solution manufacturers integrate alphanumeric displays on the front of their car PCs. See Figure 7-3.

Adding a Text Display to Your Car PC

Adding a text display to your Car PC will allow you to display information in a format that allows you to see it at a glance. Because of

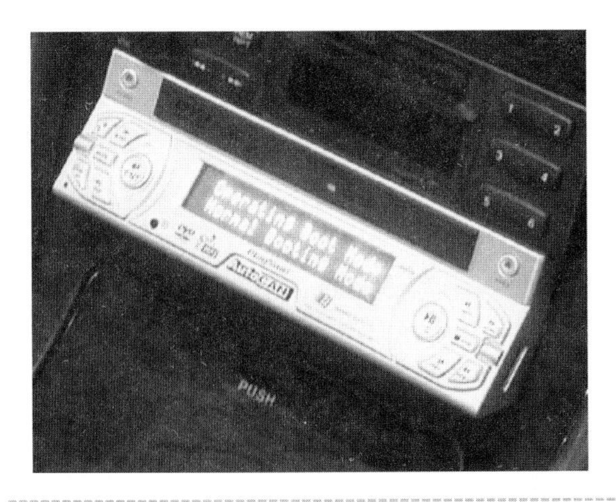

Figure 7-3 *VFD alphanumeric display in an AutoCAN car PC (Image Courtesy Digital WW)*

their small size, they can be integrated in your dashboard inconspicuously and made to appear as part of the vehicle.

There are a number of factors you will need to consider when selecting a display. One of the aesthetic concerns will be matching the LCD display's backlighting to the illumination already in your vehicle. You may want to consider whether to "invert" the display. What this means is choosing between black letters on a colored background or colored letters on a black background.

You also have a staggering choice of interface options. This will largely depend on what ports are available on your car computer. USB, serial, and parallel interfacing to LCDs can all be purchased easily.

USB devices are generally the easiest to install, as they are automatically recognized when connected—so called plug and play.

Once you have installed the hardware, you need to consider software. Displaying information from a program on your text display is done by a small piece of software called a plug-in. Manufacturers of LCD displays often provide plug-ins for common software.

Project 32: Installing a USB Text LCD

Instructions for Installing a USB Text LCD

You Will Need

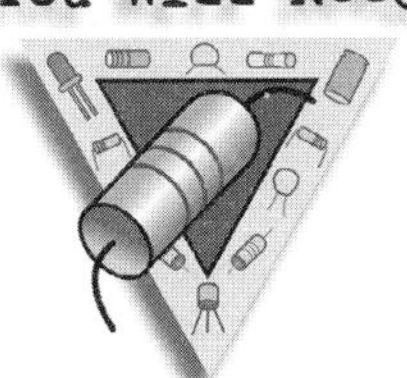

LCD

Bezel

Interface cable

Spare port

Tools

Drill and selection of small drill bits

Hacksaw blade

Epoxy resin adhesive

Masking tape

First prepare the location where you will site your LCD. If you feel confident, you can rip apart your dashboard and set it in with the clocks. But if you are less confident, you might want to mount it in a sun visor.

Consider where you are going to route the interface cable to make it inconspicuous. Will you need to remove any trim panels to gain access? Will the cable reach? If there is not enough slack in the cable to get from your desired location to your car PC, you may consider extending the cable or changing the location.

Either way, measure the size of the LCD, mark where you are going to locate it, and check that there is enough room and adequate clearance in that place to mount it.

When you are confident about all these things, prepare the area you are going to modify (destroy) by covering it with masking tape and drawing precisely the hole required.

Shade the area to be cut out.

Drill a hole in the corners of the shaded area. Now using the drill, bore a hole adjacent to the first and then adjacent to that one until you have completely followed the line of the cutout. This is a technique called *chain drilling*. When all the holes are drilled, wrap a length of masking tape around a small junior hacksaw blade. Using this end as a handle, cut the bridges between the holes. You will need to finish this area; wet and dry paper should smooth out imperfections and leave a square hole.

Push the LCD bezel through the front and affix the display behind. If no fixings are provided, the display can be glued in place using an epoxy resin adhesive.

Feed the cable behind any trim and conceal it to make a neat installation. With the car PC disconnected, plug the display into the interface cable, and then the cable into the PC.

Check all connections one last time before powering up the PC.

Power up the PC with the driver disk inside; install the driver when prompted.

Instructions for Installing a Parallel Port LCD

You Will Need

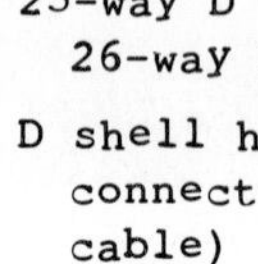

- 25-way D connector or 26-way IDC header
- D shell housing for connector (if applicable)
- Molex (PC power) connector
- Multicore cable (at least 16-way)

Tools

- Drill and selection of small drill bits
- Hacksaw blade
- Epoxy resin adhesive
- Masking tape
- Small screwdriver
- Soldering iron

The multicore cable isn't mandatory, but it makes a far nicer job of the whole affair!

To connect a parallel LCD to a parallel port, there really is little in the way of circuitry. It's just a matter of joining up the dots between the parallel connector and the LCD display.

As with every project that involves wiring, decide where the LCD is to go and look at the length of cable required.

You will need to prepare the area where the LCD is to be mounted, check that no obstructions exist and prepare the hole as detailed in Project 32 on installing a USB text display.

Most LCDs follow the same format. The HD44780 and HD44100 Hitachi chipsets are prevalent in most LCDs that are available in electronics hobby shops. It is always best to check with the manufacturer's data sheet to clarify.

Solder the connections as shown in the diagram in Figure 7-4.

You can either connect to the external parallel port connector, or if that is not available, your motherboard may have an internal header that you can use. Check out Figures 7-5 and 7-6 for more information.

Once that is done, check your work and plug the parallel port connector in first, followed by the power connector.

Take a look at some of the software in the following paragraphs. Installation instructions will be specific to each item of software.

Sites to check out:
Purveyors of LCDs
www.matrixorbital.com/
www.crystalfontz.com/
LCD Display Software
www.lcdc.cc/
http://lcdproc.omnipotent.net/index.php3
www.2morrow.com/lcdd/
www.lcdmax.de/ (German Language)
Forums
www.lcdforums.com/forums/

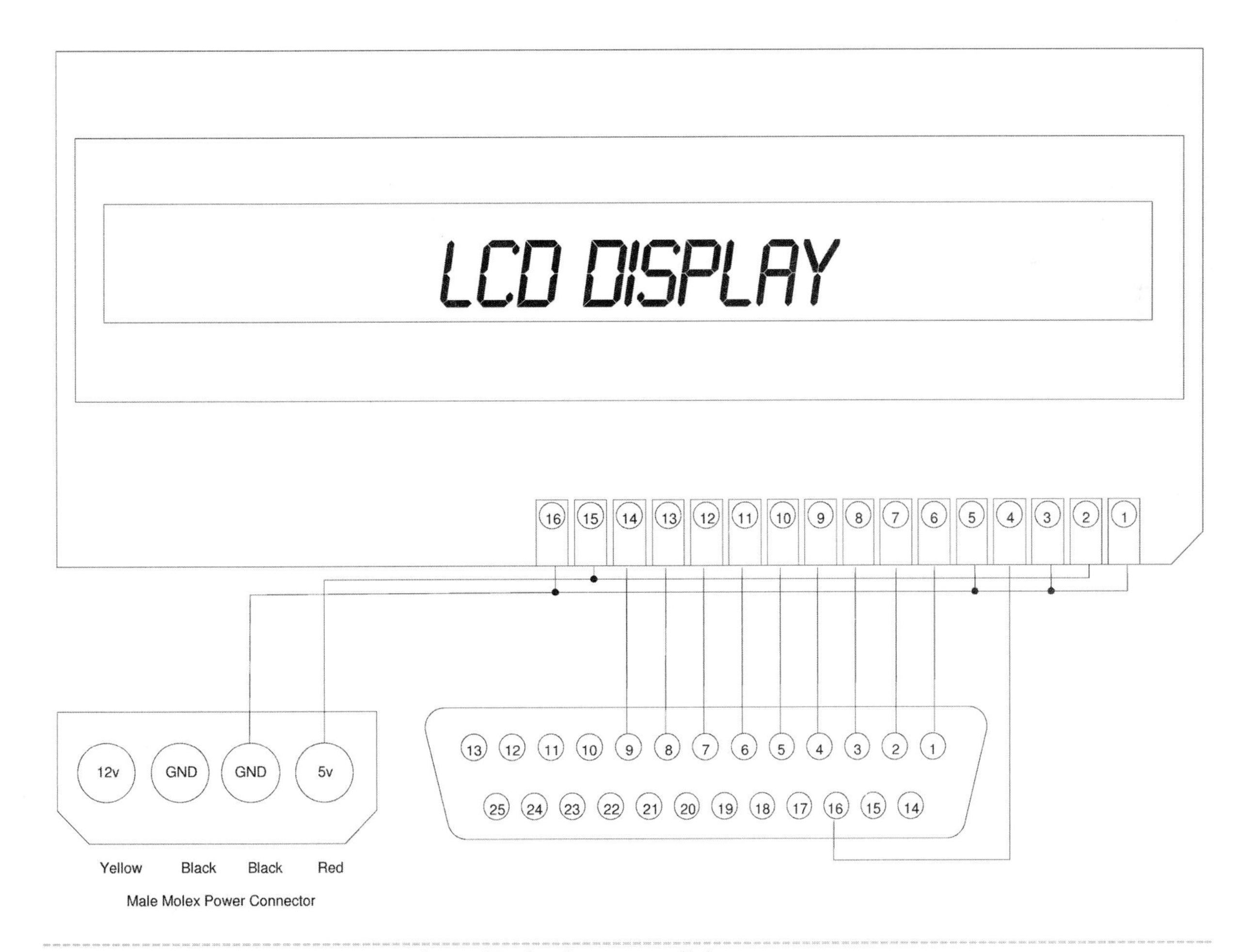

Figure 7-4 *Connection diagram for LCD display to parallel port*

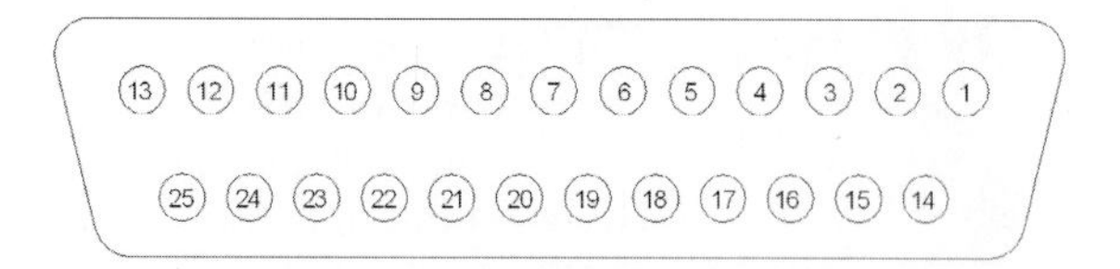

Figure 7-5 *DB25 External parallel port connector*

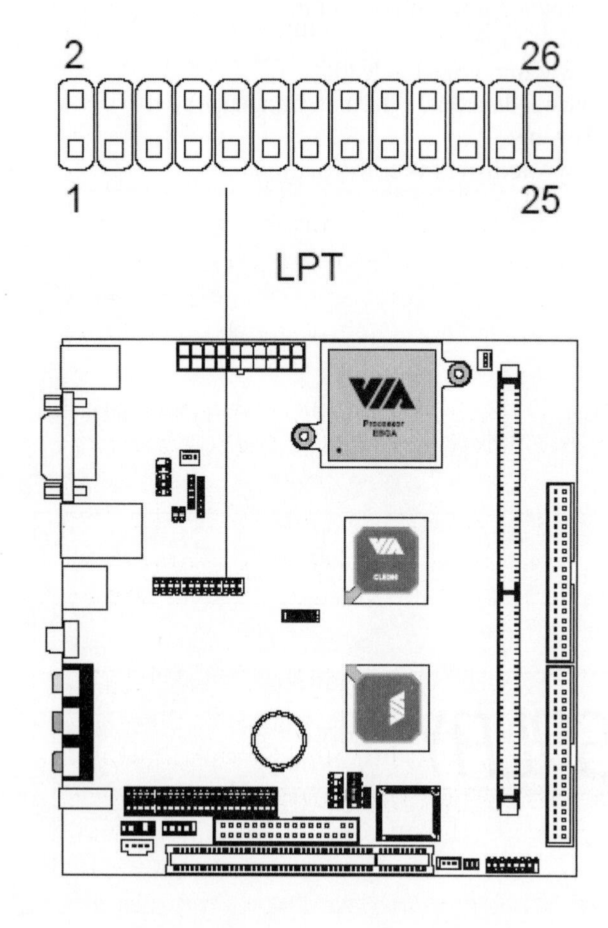

Figure 7-6 *Internal parallel port header*

Inverting Your LCD Display

You may have found that your LCD display displays black characters on a colored background while your dashboard LCD and in-car audio system display colored characters on a black background.

Changing this is no big deal. You will simply need to invert your LCD display. Inverted LCDs are illustrated in Figure 7-7.

Figure 7-7 *Normal and inverted displays*

You Will Need

You Will Need

- Polarizing filter (available from photographic suppliers such as Jessops)

Tools

Tools

- Scalpel
- Long-nose pliers
- Antistatic wrist strap

When buying the filter, you are looking for a thin flexible plastic filter. They are often used for camera work as a cheap alternative to glass filters. They may also be available in foot-square sheets for lighting work. They are sold as gelatine filters or "gels."

You need only a small area—just a little larger than your display. I found one stage lighting supplier that handed out a sampler of its lighting gels as a promotional product. Removing the polarizing filter from the filter book yielded enough filter for a 2×20 display.

Before commencing work, be sure you are grounded to dissipate any stray static charge. The CMOS devices on the LCD board can be easily damaged by static.

Depending on how your LCD display is constructed, remove the screws or clips that hold the fascia to the display. Many fascias have metal tabs that are pushed through slots in the circuit boards then twisted at 90 degrees to secure them. When untwisting tabs, do so carefully as excessive stress may cause them to break off.

Once the fascia has been removed, the polarizing film needs to be taken away from the LCD screen. Do so by slicing the adhesive carefully with a sharp scalpel.

Once this is done, lay the old piece of polarizing film on top of the new piece. Twist it until it appears black.

What is happening is that the polarizing filter will only allow light through that is polarized in a certain direction. When two polarizing filters are at 90 degrees to each other they do not allow light to pass.

Using the old polarizer as a template, cut a new one.

In the words of the great Haynes manuals, assembly is the reverse of disassembly. Of course . . . this seldom is the case!

Once your LCD is back together, connect it as it was before. The screen should now appear black with colored characters.

Full-Color LCDs

So far we have mainly focused on low-resolution text and graphics displays that offer an output similar to that found on the front of most commercially available car stereos. Although this is fine for displaying the name of the track that is playing or the temperature of your CPU, it is not good for displaying maps or viewing rich media such as DVDs.

Two types of graphical display can be used with a car PC: (1) the very low-resolution black and white LCD displays that use an interface similar to the one used in the alphanumeric displays mentioned in this section and (2) the high-resolution color displays that can be used with a VGA or composite output. Installation of a low-resolution black and white graphical display is broadly similar to the installation of the alphanumeric displays mentioned earlier in this chapter. High-resolution color displays, on the other hand, have vastly different considerations (see Figure 7-8).

When selecting a suitable screen for your car PC, it is important to consider glare, reflectance, and brightness. For your screen to be visible in moderate sunlight it is recommended that you get a screen that is 400 nits or higher. This is a good trade-off between price and performance and will prevent images from looking washed out.

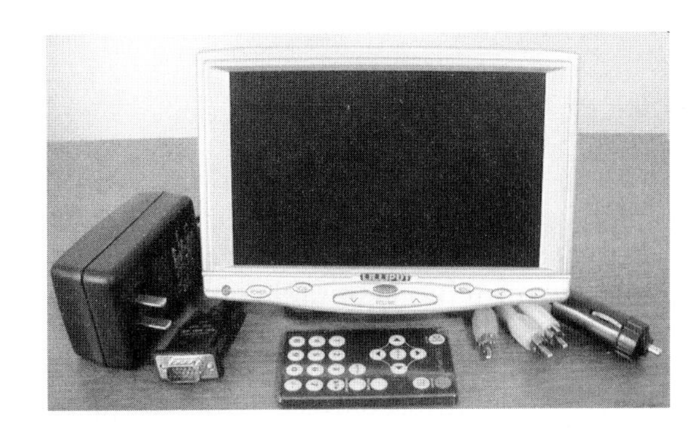

Figure 7-8 *Seven-inch widescreen color monitor*

Another consideration is that adding a touchscreen will reduce brightness slightly as the light has to pass through another layer. Furthermore, a touchscreen will increase glare by a small amount as it adds another surface for the light to reflect from.

At the time this book is going to press, no standalone transreflective screens are available that are known to me. However, if these are available in the future, they would certainly present a superior solution to a standard backlit LCD display technology.

When specifying a display for your system, it is important to consider whether you will be using a VGA or TV-OUT interface. The following are the pros and cons to using a VGA or composite (TV-OUT) display.

VGA Displays

Pros

- High resolution (suitable for use with navigation)
- Most motherboards support a VGA port

Cons

- Expensive

Composite (TV-OUT) Displays

Pros

- Cheap

Cons

- Low resolution (suitable only for video and low-resolution graphics)
- Not all motherboards or graphics cards have TV-OUT facility

Mounting Options

Once you have selected your display element, you will need to decide how you are going to mount it. A wide collection of screens is available with different mounting options. Furthermore, if you choose to go for a standard screen, hardware is available that allows you to mount it in a variety of ways in your vehicle.

Let's look at what you get with an LCD kit. Figure 7-5 illustrated the 7-inch widescreen LCD kit available from mp3Car.com.

In the kit you get the following:

- A VGA cable for connecting your car PC to the monitor
- A universal monitor stand with an adhesive base (you can stick it to your dashboard)
- A standard composite video cable
- Credit-card-size remote control with battery
- Power adaptor (mains)
- Power adaptor (12V)

If you are just starting out, then this type of monitor is just great. It offers you a great deal of flexibility and works well in or out of the car. However, many more mounting possibilities exist.

One that can be used with this standalone type monitor is the gooseneck. A gooseneck can often be seen on professional microphone stands. It is basically just a flexible rod that will allow you to support your screen in any orientation and will allow you the flexibility to bend it into a new position. An illustration of a gooseneck can be found in Figure 7-9.

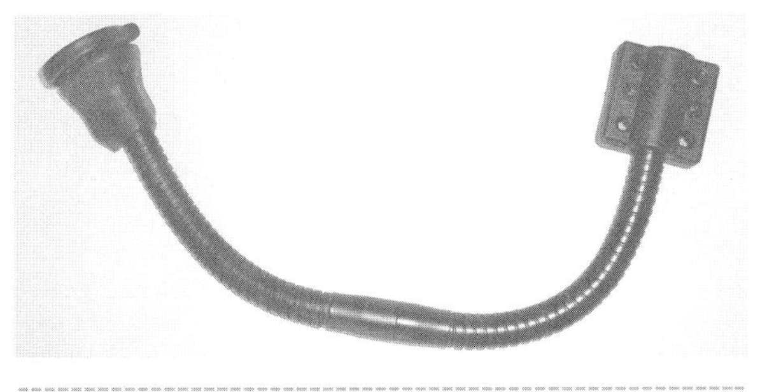

Figure 7-9 *Gooseneck for in-car mounting*

Be aware that if you buy an LCD with an integrated TV tuner, it will need to be compatible with the television standards in your country. Imported foreign LCD screens may not function correctly.

Project 33: Installing a Dashboard Flip-Out LCD

A flip-out dashboard LCD is a good way of squeezing a widescreen display into your dashboard while using the minimum amount of space.

Many aftermarket in-dash LCDs are available. However, if you already own a Lilliput/Xenarc/Gain LCD screen, then Digital WW (see Appendix A for contact information) sells a fantastic kit that allows you to convert your screen to in-dash operation.

You Will Need

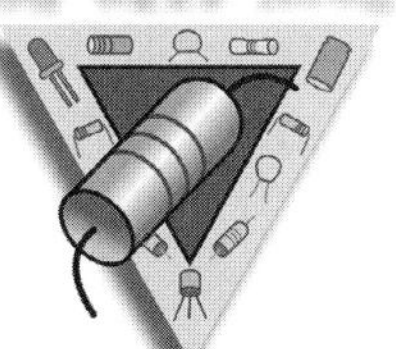

- In-dash monitor and mounting kit
- VGA cable
- Power cable
- Crimps

Tools

- DIN extractor prongs
- Coat hanger and side cutters
- Crimping tool

Installing a dashboard-mounted flip screen is much like installing any other car stereo in your dash. The hole that it all fits into conforms to a DIN standard so every DIN-size radio will fit in every DIN-size hole!

The one thing that you need to check is that there is sufficient depth in the hole to accommodate the unit. This is a simple case of measuring with a ruler!

You may find that you have a little difficulty in removing your old car stereo. Help is at hand with a DIN extractor tool. Or in a pinch you can use an old coat hanger. To modify the coat hanger, you will need to cut off the two shoulders of the hanger with a pair of side cutters so that the shoulders are about the same length. These can then be inserted in the side holes of the radio to allow it to be extracted with ease.

Fitting the new screen is simple. You will need to route the cables before connecting, if your car PC is mounted directly below in another DIN slot you will have little problem. Connect the cables for VGA and power to the back of the screen and slide it into the slot.

Because some localities do not allow the use of video screens while the vehicle is in motion, it is possible to connect a relay in the power line to disable the screen's power source when the handbrake is off. This ensures that you are complying with local legislation.

Now every time you want to use the monitor, simply press the button on the front and it will slide out.

Project 34: Installing a Flip-Down LCD

You Will Need

You Will Need

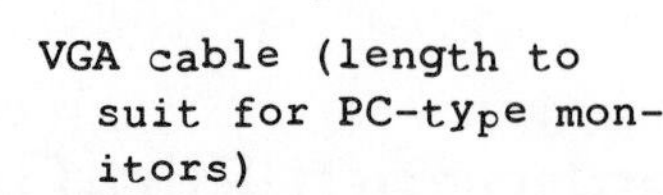

RCA cable (length to suit for composite monitors)

VGA cable (length to suit for PC-type monitors)

Assortment of crimps

Contact adhesive

Plywood square

Tools

Craft knife

Screwdriver

Contact adhesive

Applicator jigsaw (optional) side cutters

Flip-down LCDs have the advantage that they do not require any dashboard space, can be concealed when not in use, and give a sleek professional cockpit look to your vehicle. They also have the advantage that when in use, they can be enjoyed by both the front and rear passengers.

One of the main problems with installing a flip-down LCD is the lack of suitable places to mount the unit when fixing it to the roof and headlining. In many vehicles above the headlining is sound-deadening material and an air gap. And directly above this is the roof skin itself. It is not a good idea to screw into the roof skin as the screws will be clearly visible from outside the vehicle! A much more sensible approach is to attach a sheet of plywood or other material to the roof headliner and screw the unit into this.

Installation Procedure

Decide where your flip-down LCD is going to be mounted. You have a choice of locations: in the front between the sun visors or farther back between the front seats (this mounting position will mean that the screen is only for the benefit of the rear-seat passengers).

You will need to ensure that you have cables that are an appropriate length to reach these locations. The route that they take will depend largely upon the vehicle layout and construction, but allow for the cables to run under the headlining and then down the A or B pillars.

Spend some time evaluating the construction of your headliner. If it is made of a rigid plastic, you may be able to mount your monitor directly to the lining, which will be able to support its weight. If you have a fabric headlining, you will definitely need to secure the monitor to either the roof panel or the cross member if there is one.

Warning

Some cars have air ducting and vents in the roof. If your vehicle falls into this category, check that the ducts will not conflict with any monitor mounting.

Headlining Method

If you decide that your headlining will be able to support your monitor, it is simply a matter of cutting a hole for the cables, gaining access to route the cables to the desired location, and screwing the monitor to the roof. When selecting screws, a good self-tapping screw will bite into the plastic. Be sure the screws are not so long that they damage your roof panel, as any damage will be visible on the outside of the vehicle!

In terms of electrical connection, most monitors require a permanent 12-volt feed, a ground, and a feed from your dome light. If you are mounting the monitor at the front of your vehicle, you will find that the dome light wiring is easy to access.

Roof Panel Method

To remove your headliner, you are going to need to remove all the rubber around your doors that secure the headliner, any overhead lights switches, or any accessories. Your interior mirror and sun visors will also need to be unscrewed to allow you to gain access to the roof.

A piece of thick plywood, cut to the size of your screen's base, can then be coated with a contact adhesive and pressed onto the roof headliner. You should hold it until the glue has formed a bond and then allow it to dry fully before proceeding any further.

If your screen comes with a separate infrared transmitter or receiver, you will need to find a mounting location for this as well. Think about where you will be using a remote control to operate the screen, and choose a location that is in the line of sight of the remote.

While the glue is setting, you might like to consider examining your overhead dome light in a little more detail.

Many overhead LCD screens incorporate a replacement or additional dome light. If you are putting this LCD monitor in the place of your old dome light, you will need to remove the wires leading to your existing dome light and connect them to the new leads on the new LCD screen dome light. If you are going to be using this screen as an additional LCD monitor, you will need to solder wires onto your existing dome light and piggyback them onto your new screen. The dome lights are wired in a parallel circuit.

If your vehicle has a sunroof, you will need to be very careful when installing your overhead screen that the monitor's wiring and fixings do not conflict with your sunroof's mechanism.

Project 35: Installing LCDs in Your Headrests

Front-seat headrests provide an ideal mounting location for screens for rear-seat passengers. They are convenient in that they are at eye level and therefore do not require passengers to strain to see the picture. Installing a headrest LCD requires modification of your existing headrests. The following project will talk you through this procedure.

You Will Need

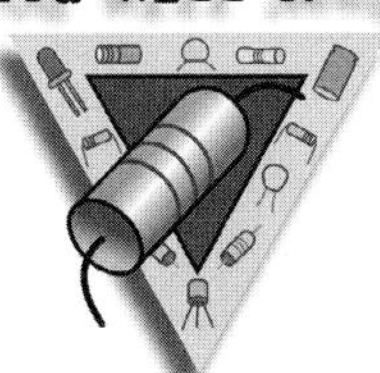

- LCD monitor
- RCA cable (length to suit for composite monitors)
- VGA cable (length to suit for VGA monitors)
- Thick velcro strips

Tools

G clamp

Scalpel or craft, Xacto, utility, or Stanley knife

Biro, ballpoint pen, or permanent marker

Remove the headrests from the seat. The headrests are typcially mounted by a pole (or two) running inside a hollow tube in the seat-back. There may be some kind of lock to prevent the headrest from being fully removed without intervention. This will need to be removed if the headrests are to be fully removed.

Square up the screen to the headrest and work out where the monitor is going to be mounted. You may want to work out where the center of the headrest is to be sure the installation is visually pleasing.

Now flip the headrest around so you are looking at the protruding poles that fasten the headrest to the seat. Almost universally these will be secured to the headrest using some kind of mounting plate, which serves a double purpose in covering the open fabric seam at the bottom. This may be attached to the headrest body using a variety of fixings—often self-tapping screws. This plate needs to be removed in order for you to gain access to the headrest (see Figure 7-10).

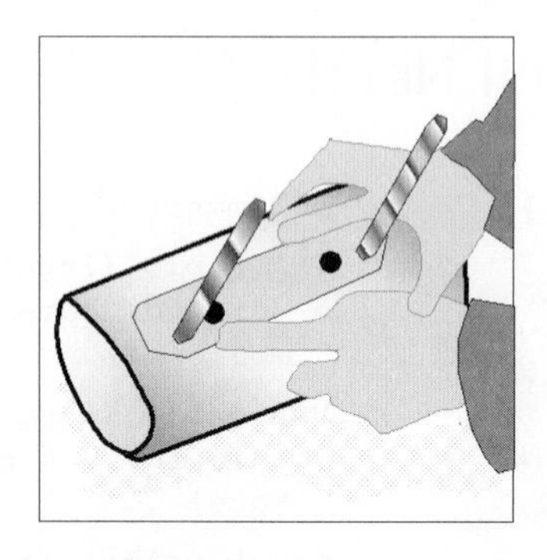

Figure 7-10 *Removing the plate in the headrest*

Once this plate is removed it will become much easier to remove the fabric from the headrest. The headrest fabric is commonly secured by zippers and clips.

The fabric should now be carefully removed from the headrest, taking care not to damage or rip the cover, as repairs to upholstery can be very costly.

You will now need to make a template of your LCD screen. Fabricate it from *medium density fiberboard* (MDF) or plywood, and make it the same height, width, and depth as your LCD screen. If you cannot obtain a piece of MDF or plywood of a suitable depth, you can laminate a couple of layers together using a good contact adhesive and G clamp.

When you have cut this template, you can use it to begin cutting the foam of the headrest. Finding the center of the foam, place the template and mark around the edge of it using a permanent marker.

You will now need to remove the foam, layer by layer, until you achieve the same depth of foam cutaway as the depth of your monitor. Next place your template snugly in the hole (see Figure 7-11).

Once you have cut the foam to a suitable depth, you will need to cut a hole in the fabric cover to accommodate your monitor. Turn the fabric cover inside-out so you can mark on the back or inside of the fabric. Do not use a permanent marker or any other pen that will bleed through to the outside of the fabric, the side you see. Instead use a biro or similar marker. Mark out a rectangle the same shape

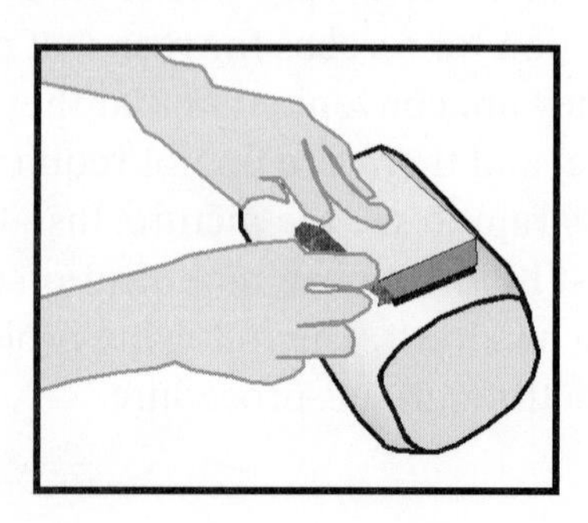

Figure 7-11 *Placing the template*

as your monitor. But remember to reverse the template as you are working on the rear side of the fabric.

Do not cut out the rectangle of fabric. Instead mark diagonals (from one corner to the opposite corner) of the rectangle and then cut along these lines with your knife.

You will now need to pull the cover back onto the headrest and trim the fabric into the hole. Don't worry if you can see a little foam, this will be covered by the LCD screen. Now line the back of the cutout with self-adhesive velcro strips. Similarly, the rear of your monitor should be covered with velcro.

Ease the monitor cabling into the hole and push it through the foam until it protrudes near one of the headrest supports. You can now ease the monitor into the hole. It should seat firmly and affix to the headrest with the velcro.

You may need to cut the securing plate to accommodate the new cable.

You can now reassemble everything, ensuring that the securing plate is used to hold the fabric taut to the housing. You will need to run the monitor cables inside your seat. See some of the earlier tutorials in this book for instructions on how to take a front seat apart.

Chapter Eight

Input Devices

Using a car PC is a completely different experience from using a standard desktop PC.

As little space is available inside a vehicle, the way the user interfaces with the car PC need to be adapted to the environment in which the user is working. The fact that the user may well be the driver means that the input devices need to require minimal input from the user, who will be probably be occupied.

Using a desktop PC is a pretty resource-intensive activity. If you are trying to type, you generally use two hands, and when browsing the web, you may keep one hand on the keyboard and one on the mouse. Even using a laptop, your hands are occupied most of the time.

With a car PC, the driver needs his or her hands for driving, with possibly a minimal input to the car PC system. Think about using a PC as an audio system compared to the experience of using a car stereo. To turn the volume up in your favorite media player requires a little effort on the part of the user in a touchscreen. A touchscreen is probably the most intuitive and easy-to-use interface, and it is versatile.

Project 36: Installing a Touchscreen

You Will Need Touchscreen kit to suit your monitor

Touchscreen kits are available from www.digitalww.com; touchscreen kits are available in the following sizes:

- 5.6-inch touchscreen kit
- 7-inch touchscreen kit
- 7-inch Lilliput touch panel
- 8-inch touchscreen kit
- 10.4-inch touchscreen kit
- 12.1-inch touchscreen kit
- 15-inch touchscreen kit
- 17-inch touchscreen kit

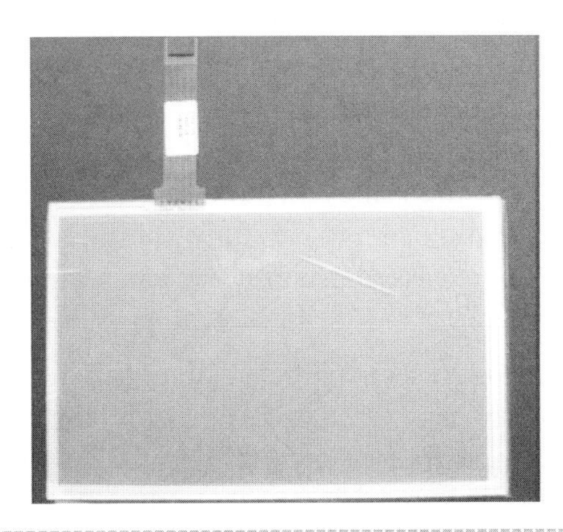

Figure 8-1 *Touchscreen panel (Courtesy Digital WW.com)*

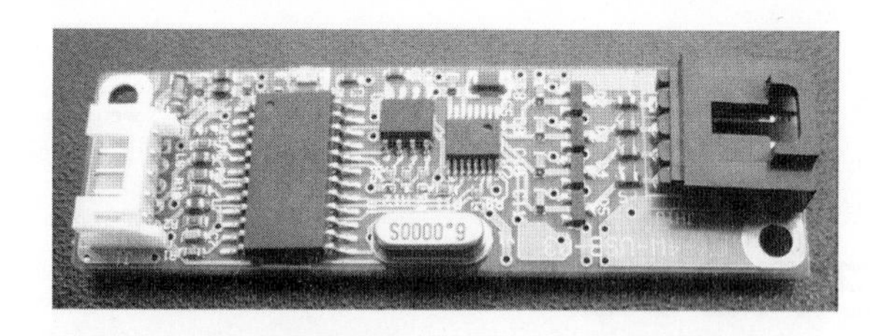

Figure 8-2 *Touchscreen controller (Courtesy DigitalWW.com)*

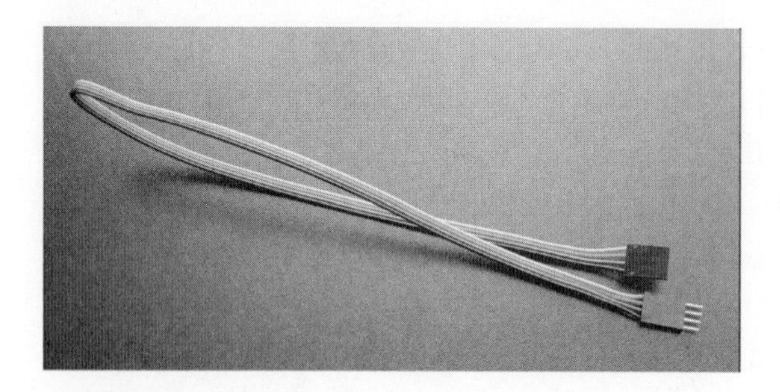

Figure 8-3 *Controller-to-panel cable (Courtesy DigitalWW.com)*

The kit will contain a touchscreen panel (see Figure 8-1), a touchscreen controller (see Figure 8-2), a controller-to-panel cable (see Figure 8-3), and a USB cable.

Following these instructions will allow you to install a touchscreen in most makes of an LCD.

Consider the following when buying a touchscreen kit. Try to find one where the "active area" most closely matches the visible area of your screen. Invariably, the touchscreen controllers have a USB interface.

Unscrew your LCD monitor carefully. Exact details will vary with the make and model of the screen. Usually the screws employ a small Philips or torx head. Once all the screws are removed, put them safely to one side and begin to carefully pry the two parts of the case apart until you can see inside the case.

As a rule, there will be a ribbon cable and/or series of small cables that join the circuit boards in both parts of the case (see Figures 8-4 and 8-5). These will usually run between the main board inside the LCD display and the screen, any buttons on the front of the case, and any socket on the rear. A speaker may also be present. All of these can usually be disconnected at one end. With the exception of the cable connecting the main board to the LCD panel, disconnect them noting the orientation of the plug and position.

Now it is time to remove the LCD panel from its casing. Again designs between manufacturers vary, but quite often the panel is affixed to the casing using a variety of screws and plastic lugs. If plastic lugs are troublesome, they may need to be carefully removed using side cutters or snips (see Figure 8-6).

Now try fitting the touchscreen against the frame and note any discrepancy between the dimensions of the front casing and the active area of the screen. If the active area of the touchscreen is larger than the opening in the front casing, it will need to be spaced to allow the touchscreen to function correctly.

Figure 8-4 *Disconnection of backlight (Courtesy DigitalWW.com)*

Figure 8-5 *Disconnection of ribbon cable (Courtesy DigitalWW.com)*

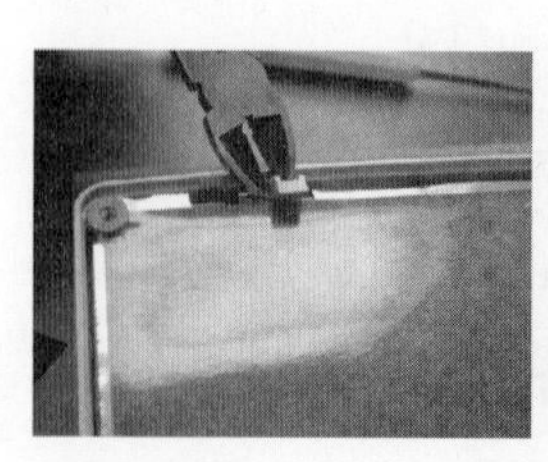

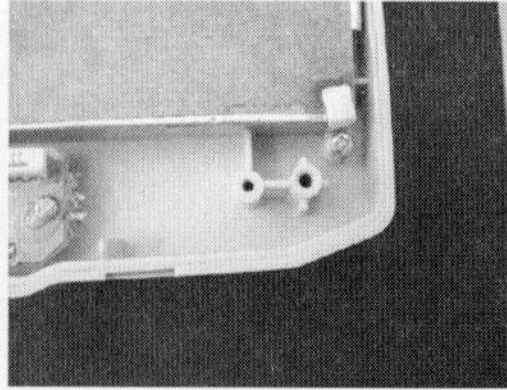

Figure 8-6 *Removal of plastic lugs (Courtesy DigitalWW.com)*

Figure 8-7 *Cardboard gasket (Courtesy DigitalWW.com)*

A number of methods can be used to achieve this. One method is fabricating a cardboard gasket (see Figure 8-7). Measure the active area of the touchscreen and draw a rectangle a few millimeters larger on a sheet of thin card. Now draw a frame that is a centimeter or so larger than the first rectangle (this may need to be adjusted depending on the molding of the front plastic case). This will prevent the touchscreen from contacting the case and causing the pointer to go off course. Double-sided tape can then be used to hold it in place.

Another method is to run a small bead of silicone sealant around a rectangle that is larger than the active area of the screen. It must dry *fully* before the touchscreen gets anywhere near it!

Thin cork tiles are another possible material that can be used for a gasket, although I have found that the extra thickness often makes it difficult to shut the case fully when the screen is installed.

Foam draft excluder tape is one solution that is practical and not very messy. Either run the strip around a rectangle larger than the active area or cut out small strips to space the screen.

Before proceeding further *thoroughly* clean both the LCD panel and touchscreen panel using an antistatic monitor cleaner. Any dust will be there forever so make sure you remove it all. Kensingtons makes some good screen wipes. If you use a spray and cloth, make sure it is lint free. The last thing you want is to have to look at fabric fibers when trying to watch the latest DVD!

Next put the touchscreen in place with the LCD panel on top. You may need a little bit of out-of-sight double-sided tape to make sure that it all holds. Screw the controller board on top of the LCD and touchscreen. This should hold everything in place.

Now, remembering the orientation of the cables, connect the buttons, sockets, and speaker (where fitted) to the mainboard. Assembly is the reverse of disassembly as the Haynes manuals state.

Closing the two halves together, see if a suitable place can be found for the touchscreen controller. If it can be mounted within the enclosure, connect it to the touchscreen cable and drill a suitable hole in the rear casing for the USB cable. You may find that the USB cable can be routed through a speaker hole or vent hole.

Next screw the casing together. If the controller will fit inside, make sure the USB cable is hanging out. If it won't fit, make sure the screen ribbon cable is hanging loose.

If the controller is mounted externally, you may want to think about protecting it. A film can epoxied onto the rear of the case should be adequate, although more elegant solutions are possible with a little time and imagination.

Once the screen has been reassembled, connect it to the car computer using the USB interface for the touchscreen, and the VGA for the monitor and power (derived from either the mains or 12-volt supply).

Upon powering up the PC, the touchscreen should be identified as new hardware. Install the drivers from the disk provided.

The next step will be to calibrate the screen. This will vary depending on the software provided. The usual sequence is to tap a series of points (indicated by a crosshair on the screen).

Try to be as accurate as possible when touching the points as this will have a big impact on the accuracy and usability of the screen. If you can use a PDA stylus, great; if not, try to use a fairly sharp (but not too sharp) point such as the cap of a ballpoint pen. Don't use a finger for the initial calibration as it will produce inaccurate results.

Once the screen is calibrated, that's about it! You may be required to restart your car computer once the drivers are installed, but upon the reboot, the screen should be fully functional.

Troubleshooting

If there is no response from the screen at all, power everything down, disassemble the monitor, and check all connections. Make sure that the screen is connected to the controller properly. If the car computer recognized the USB device, the chances are the USB connection is fine.

If the LCD display is working intermittently, check that it is connected properly to the power supply. It is quite possible that a cable has been inserted incorrectly upon reassembly.

If the cursor continually drifts toward one corner of the screen when you move your finger vertically, it means one of the wires in the touchscreen is broken. If the break in the wire can be identified, then it may be possible to repair it.

If the cursor does not follow your finger exactly when you touch the screen, you can usually fool the calibration routine. Tapping the screen at a point that is offset from the crosshair by a distance proportional to the error should correct the problem in most cases. If after several attempts at recalibration there is little success, the problem may be a result of the EEPROM chip used to store the calibration data. In this eventuality a replacement board should be sought.

If the touchscreen is new, it may be possible to arrange an exchange or replacement. If it is not possible to obtain a replacement, conductive paint that is used to repair broken rear-window heaters could be used to fix a small break in the track. To do this, mask off the area and apply the paint to the broken trace. Allow the paint to dry and then peel off the masking tape. This may solve the problem.

Project 37: Installing a Rotary Controller

A rotary controller, although simple in operation, can be one of the most direct ways of adjusting presets for such things as volume control. It makes much more sense to be able to turn a knob in proportion to the amount of change required rather than pressing a button continuously, which can prove distracting.

One such controller is the Griffin PowerMate (see Figure 8-8), a sleek aluminum knob that connects to the PC using a USB interface. It has a number of modes of operation. In addition to being able to turn it left and right, additional control can be exercised by pressing the knob in and by pressing and turning simultaneously.

My only criticism of this device is the rather short nature of the USB lead. For a trunk-mounted PC, this could prove problematic. But extension of the cable using screened, four-core cable is quite simple and does not appear to adversely affect operation. It does, however, spoil the aesthetic of the device somewhat.

For mounting on a dashboard, a highly recommended modification is to remove the side-mounted cable grommet, drill a hole through the plastic base, and reroute the cable so it exits through the rear rather than the side. The PowerMate can then be fixed onto the dash by drilling a hole to conceal the cable and using an epoxy resin glue to mount the device securely.

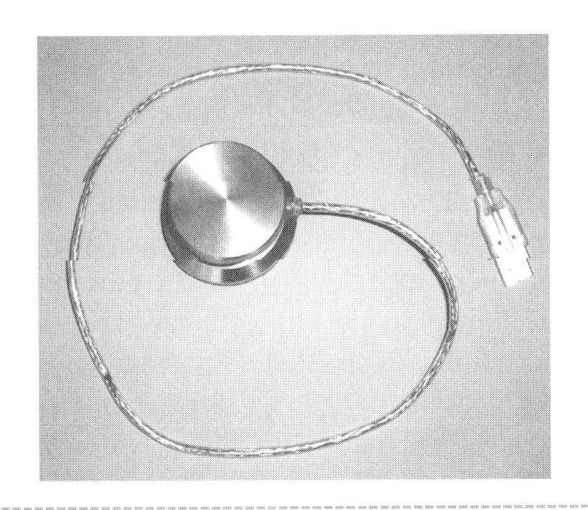

Figure 8-8 *Griffin PowerMate USB device*

The drawback to this method is that it leaves a hole where the grommet was once present. Without machining a new enclosure for the device, this cannot be avoided. So the best course of action must be to mount it with the hole facing down in order to make it as unobtrusive as possible.

Software installation is easy. Once plugged into the USB port, insert the CD and autorun should start.

Troubleshooting

Manual Installation

Under certain circumstances, the device will not install automatically. The result may be that the cursor moves left and right when the knob is rotated but does nothing else. In this case the drivers may need to be installed manually.

Go to Start > Control Panel > System > Hardware > Device Manager. On the device manager list, try to locate Griffin PowerMate. It will be found in either Human Interface Devices, Mice and Other Pointing Devices, or Universal Serial Bus Controllers. Once you have located the device, go to the Properties window and select the Driver tab. Then click

Update Driver. Rather than follow the automatic installation path, follow the manual one by selecting Have Disk when the option presents itself.

Select Browse and, where X is the letter of your hard disk, look in:\Program Files\Griffin Technology\PowerMate\Driver. The name of the driver file is powrmate.inf. Now reboot the PC.

Q. I am running a media player in the background while using navigation software, but my PowerMate won't let me readjust the volume.

A. PowerMate will only work with the active application.

Windows 2000 Users

Windows 2000 users should be sure they are using at least Service Pack 3 to support the PowerMate.

Project 38: Controlling Your Car Computer Using a Mobile Phone

You Will Need

You Will Need

- Ericsson T68m, T68i, or T610 mobile phone
- Bluetooth adapter

Tools

- Phone Front software available from www.phonefront.com/

Phone Front allows you to use your mobile phone as a remote control for your car PC.

To enable you to use Phone Front, you will need a connection between your mobile phone and your car PC. One of the following methods will suffice: infrared, serial, or Bluetooth (although Bluetooth is to be preferred where available). Some of these terms are discussed in Chapter 14 on wireless connections.

Installation Procedure

This project is targeted at users who will be employing a Bluetooth interface, which is the preferred method where possible.

You will need to install your Bluetooth adaptor in accordance with the manufacturer's guidelines. Once this is done, you must partner your mobile phone with your car PC. Bluetooth should be enabled permanently on your mobile phone for best operation. On the Ericsson T68i, the way to do this is by using the phone joystick and navigating to the following:

Connect (using the menu and joystick)

Bluetooth (which is option 4)

Options (which is option 4)

Operation Mode (which is Option 1)

On (which is selected using the radio buttons)

To install the software successfully, you will need to have Microsoft's .NET framework installed on your car PC. This is available from www.microsoft.com.

Phone Front Configuration

When you start the Phone Front application, you will notice a new icon in the taskbar, next to where your MSN Messenger logo would be. If you double-click this icon, you will bring up the main menu. You need to select the Configure option. Next select Select.

You will now need to select the phone. If using the latest Microsoft drivers, this can be done by selecting the device name directly. Failing this, if you are using older drivers, you will need to select the COM port assigned by the legacy driver.

What follows this step is a short verification procedure in the software used to obtain a license. You will need to have an active internet connection to register the software.

Using this software will in no way disrupt the normal functioning of your mobile phone while you are using it. You will still be able to receive calls and *short message service* (SMS) messages.

To use the software, you will have to install one of the scripts, which add additional menu's to your mobile phone to allow you to control your favorite PC applications. One of the modes that the software allows is tracking your desktop. This is where your mobile phone is in control of the active window. This can be disabled if required.

Project 39: Controlling Your Car PC Using a PDA

You Will Need

WiFi-enabled pocket PC compatible PDA

WiFi-enabled car PC

Tools

NetOp software

NetOp, a fantastic little program from www.netop.com, was originally designed to allow IT professionals to control PCs remotely over a network for administration and maintenance. By using the software ingeniously with a PDA, it allows you to control your car PC over a network, albeit a wireless one!

NetOp allows you to view the desktop of your car PC through your PDA and control it using the touchscreen. This presents us with some fantastic opportunities! Does anyone remember the James Bond film where bond drives his vehicle from his handheld PDA? Well, you may not be able to drive your vehicle, but you can certainly change a song or turn up the volume!

NetOp works by sending a copy of your car PC desktop to your PDA. You can either allow the view to fill the screen or, alternatively, you can scroll around the view using the stylus. You then interact with the screen just as if you were using your car PC touchscreen, only wirelessly! The information is all sent wirelessly as standard network traffic.

To install NetOp you will first need to install the device drivers for your WiFi hardware on both the car PC and your pocket PC. It is then a matter of installing the NetOp software on both car PC and pocket PC. Installation will differ slightly depending on your pocket PC model.

Chapter Nine

In-Car Audio

For a car computer to be truly effective, it needs to be interfaced to the audio system of the vehicle in question. It is no fun watching a DVD without sound. Neither is it any good looking at a map without audible directions.

For some installations, where an existing radio cassette player is left in place, it is often easier to use a tape adapter (see Figure 9-1) or FM modulator.

The disadvantages of this approach are that the installation appears to be messy (with a wire trailing from the existing HiFi), and the sound quality is poor compared to that from a hard-wired installation, as the audio signal is transmitted by induction between the transmitting tape head (enclosed in the dummy tape) and the receiving tape head (in the car audio system). This step of the operation introduces electrical noise into the system and is far from ideal.

Tape Adapters

The beauty of tape adapters is that they allow your PC to be used with any vehicle with a radio and cassette player. This can be useful when testing and troubleshooting the car computer, as it allows quick and easy connection for diagnostic purposes.

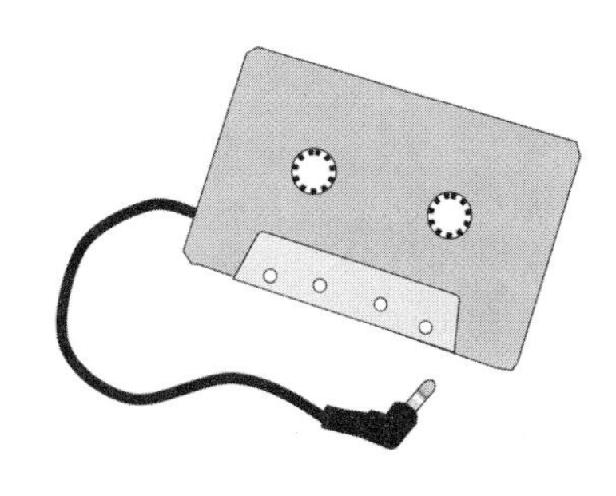

Figure 9-1 *A tape adapter*

Project 40: Constructing a Tape Adapter

You Will Need

You Will Need

- Old cassette tape
- Tape playback head
- Length of screened cable (to suit)
- 3.5mm stereo jack (or connectors to suit)

Tools

- Small jeweler's screw-drivers
- Side cutters

Constructing a tape adapter is a relatively simple affair. As can be seen from Figure 9-2, the wiring is relatively simple.

Purchase a spare tape head from your local electronics store and find an old spare audio cassette.

Take the audio cassette apart carefully, retaining the screws you remove. Much of what is inside the cassette can now be discarded. You have no further use for the magnetic tape itself, although the plastic spoked wheels should be retained and left in place as they provide a guide when inserting the tape cassette into your player. And the small pieces of plastic associated with guiding your tape through the bottom of the cassette can now be thrown away as they serve no useful purpose.

Drill a small hole in one of the corners of the tape cassette. Attach a length of double-core screened cable using a dab of glue to act as a strain relief.

The screened cable must now be soldered onto the contacts on top of the tape head. You now need to place the tape head in a position where it will align with the tape player's playing head when the cassette is in the player. This will be in the center of the cassette, roughly where the small sponge square was that supported the magnetic tape previously.

Some of the plastic inside the cassette may need to be removed to accommodate the tape head. The tape head can be secured with a little glue. Once all of this is done, the tape head should be screwed back together and the glue allowed to dry.

Once everything is set in place, the jack plug can be soldered onto the screened cable and the tape adapter is ready for use.

FM Modulators

FM modulators literally give you your own radio station in a car.

An audio signal is fed into the FM modulator, which transmits it at low power on an FM station of your choice. This enables you to set a preset station for car PC so that when you want to listen to voice navigation instructions or music you press the allocated preset on your stereo. Your car stereo now plays the sounds coming from your car PC as if they were simply another radio station.

The advantage with this approach is that no hideous trailing cables are visible, as the circuit can be concealed and hidden alongside the car PC.

The disadvantages with this approach are that in many localities this practice is not legal as it interferes with other people's audio equipment. Also, legal issues exist with transmitting audio if you are not the copyright owner.

Again, in terms of audio quality, the signal is not as clear as it would be with a direct connection, and some signal will be lost in modulation and transmission.

In Figure 9-3 you can see the stereo FM modulator available on mp3car.com.

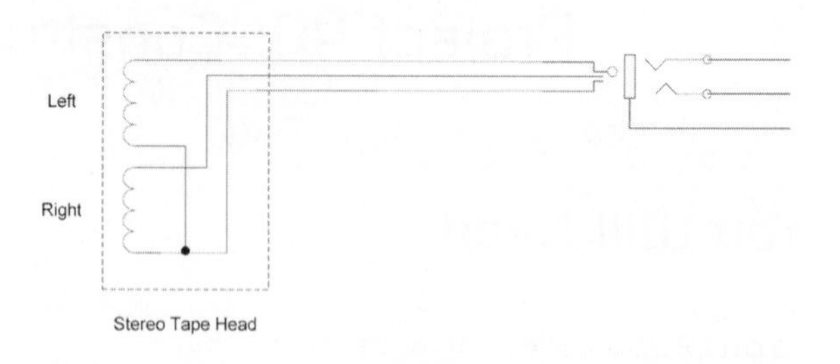

Figure 9-2 *Tape adapter wiring diagram*

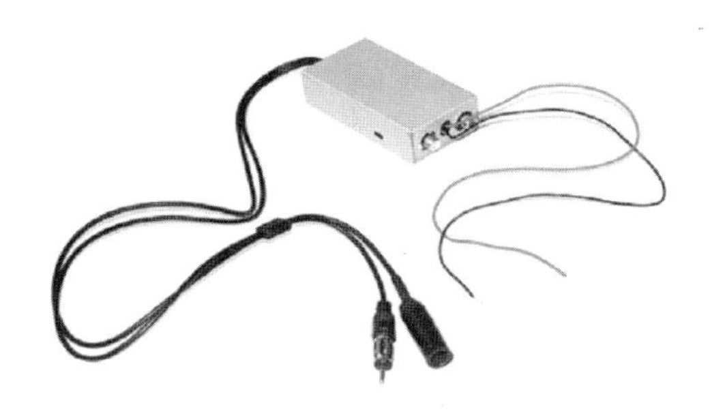

Figure 9-3 *Commercially available stereo FM modulator (Courtesy Digital WW)*

Project 41: Constructing an FM Modulator

You Will Need

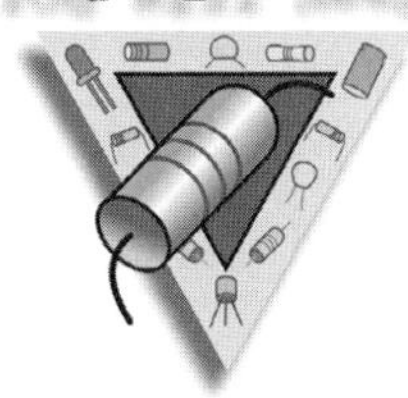

- Resistor 100R (×2)
- Resistor 1K
- Resistor 10K (×3)
- Resistor 100K
- Resistor 1M
- Preset resistor 22K
- Capacitor 0.1 μF (×3)
- Capacitor 4.7 μF
- Capacitor 10 μF
- Variable capacitor 4 to 40 pF
- Inductor 0.1 μH
- Transistors 2N3904 (×2)
- Hookup wire (for aerial)

Tools

- Side cutters
- Soldering iron

The downside of this circuit is that it will transmit only mono audio. For high-end audio fans, this is not the hot ticket. But to get you up and running, this circuit is to be highly recommended as it costs next to nothing to construct and uses no specialized components or ICs.

Where the car computer is being used for applications like GPS navigation, this circuit is a great way of sending the sound without unsightly wires. For HiFi audio or multimedia, direct-wired connection is far more preferable (see Figure 9-4).

Stick to using ceramic capacitors; nothing special is required. However try to site them somewhere away from the heater or other objects that may cause temperature change, as changes in temperature will affect the operation of the circuit.

The inductor can easily be made by wrapping some 22-gauge wire around a ballpoint pen 10 times.

The antenna is 12 inches or so of ordinary hookup wire.

The variable resistor is adjusted to attenuate the incoming line signal to a suitable level.

The variable capacitor provides a method of tuning the circuit.

When adjusting the variable capacitor, use a trimmer tool (which is plastic and nonconductive).

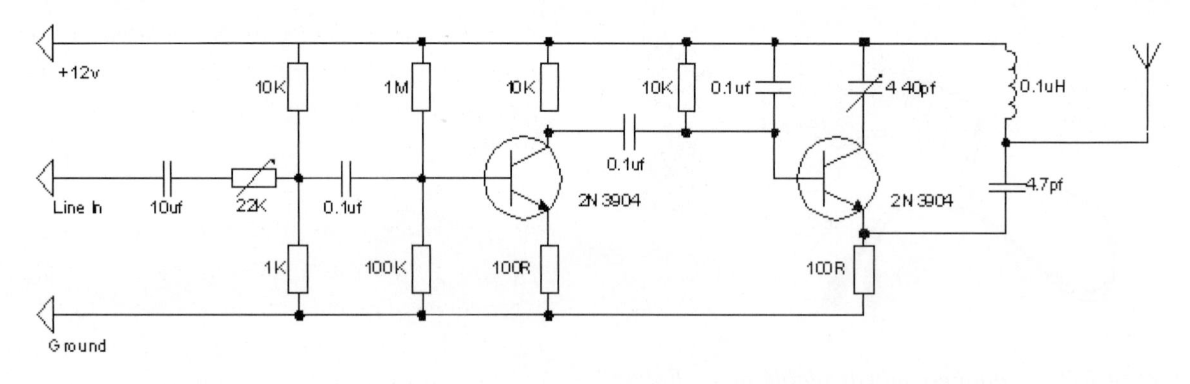

Figure 9-4 *FM transmitter schematic*

Project 42: Connecting to Auxiliary Inputs

This method is to be preferred when it is available.

Many car stereos now come with an *auxiliary* (aux.) input for connection of a CD player or the like. This can often be exploited by a car PC; it is a simple affair to make a lead with two phono plugs on one end (for left and right channels) and a 3.5mm jack on the other. The connections are illustrated in Figure 9-5.

The advantages of this method are that the sound quality is far superior to the previous two methods. If the inputs are located on the rear, it can also be a very neat method of installation.

The disadvantages are that not all car audio equipment supports this feature so you may need to upgrade your system.

Volume control can be adjusted on either the car stereo or the car PC. Probably the most user-friendly option is to set a sensible output level on the car PC and adjust the volume routinely using the car stereo head unit. The whole setup is illustrated in Figure 9-6.

Connecting Directly to an Amplifier

A more elegant solution, if all of your audio needs are met by your car PC, is to connect directly to your amplifier. The advantages of this approach is the simplicity with which it can be achieved and the superior sound quality it has over some other methods, as the path from car PC to speaker is more direct without other components in between (see Figure 9-7).

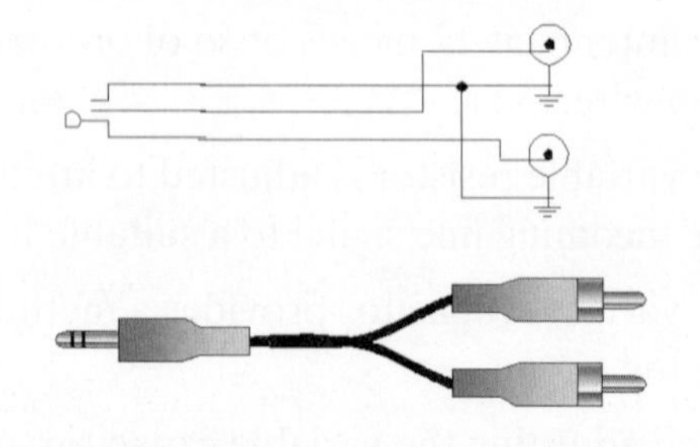

Figure 9-5 *Line out, head unit auxiliary connection*

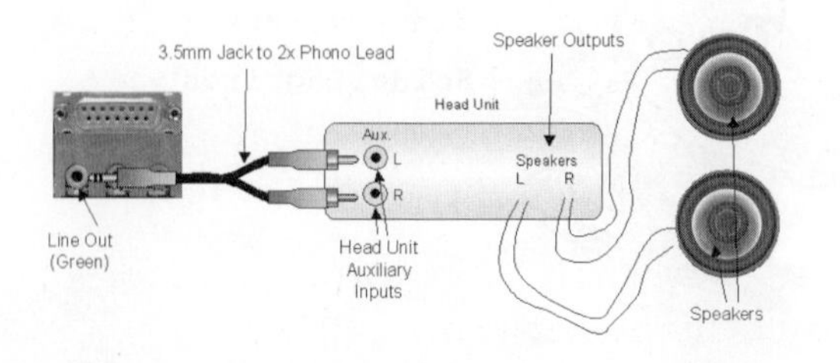

Figure 9-6 *Connecting your car PC through a head unit*

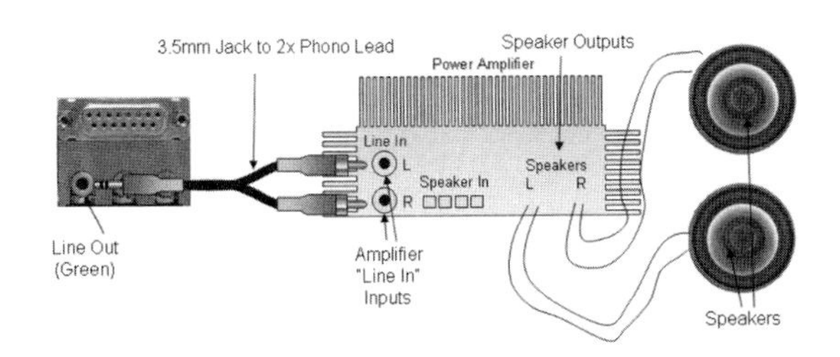

Figure 9-7 *Connecting your car PC directly to an amplifier*

Most car-audio amplifiers have two sets of inputs. In most cases, one input will be in the form of a single phono socket per channel. This input is a line-level input. Connections will normally be made using a screened cable terminated in a phono plug. The signals fed into this input are low level. This is the set of inputs you need to use for the car PC.

The other set of inputs will normally be in the form of a series of screw connectors, similar to those used for the outputs for the speakers. These are high-level inputs and are intended to handle the output at speaker level from the head unit. Your motherboard will not produce high enough signal levels to drive this input.

In the vast majority of cases, only one set of inputs should be used unless your amplifier specifically states that it can use two sets of inputs.

In addition to the previously mentioned advantages, further advantages to this method are that if you want in-car surround sound, it is far easier to procure in-car amplifiers with 5.1 capability than it is to find a head unit with 5.1 auxiliary inputs.

The disadvantages to this method are the fact that you are unable to use a radio and cassette player with a car PC, which may or may not be an issue, and the fact that unless you integrate a suitable device into your car PC, you will lose radio functionality.

Connections for surround sound are similar to those for stereo, with the exception that there are more channels for surround sound.

Volume control is now solely adjusted via the car PC. The amplifiers' gain should be set to a sensible level so that no distortion exists when the volume control on the car PC is set to maximum levels.

Project 43: Setting Up an In-Car Surround Sound

Even the most basic built-in motherboard audio seems to come with four-channel outputs—even 5.1 surround sound. For more demanding users, sound cards are available with 7.1 surround sound.

The one problem here is that most car audio systems are not geared to multichannel audio. This means that either multiple amplifiers or an expensive, dedicated in-car surround amplifier is required.

With the growth of in-car multimedia, 5.1 amplifiers are becoming quite common. However, if you want to pay for the more advanced 7.1 option, you will need to include an additional stereo amplifier in the setup. This should be chosen so that it has a similar power output to that of the amplifier used for the other channels.

With some on-board audio systems, using the motherboard in surround mode means sacrificing the mic and line-in inputs. If voice control is being considered, this will not be a viable option. Consult your motherboard user's manual. You may find a duplication of the motherboard's audio connectors on the board in the form of a header, in which case you will not lose mic and line-in functionality.

When using Mini-ITX boards, an important consideration is whether the addition of a sound card warrants sacrificing (often the only) *peripheral component interconnect* (PCI) slot if there is already sound card functionality on the motherboard. In many situations, an external sound card is a very attractive option.

Using an external sound card saves using PCI slots, which are scarce on a small motherboard. On small form-factor motherboards, there is often only one PCI slot.

It allows the sound card to be mounted remotely from the car PC, keeping the audio signal lead lengths short between the sound source and the amplifier.

As audio is transmitted digitally via a USB interface between the motherboard and sound card, little degradation of signal occurs despite long lead lengths. This is because digital protocols are robust and not so susceptible to interference, whereas analogue audio signals are prone to electrical interference. In a car environment, a lot of electrical noise exists, so anything that will improve audio quality is to be welcomed.

Once an audio source is chosen, the next big question is how to arrange the speakers for correct staging and positioning of sound. The choice is one of personal preference, and in an automotive environment many trade-offs can be made. Positioning of speakers is hard enough in a home environment; in an automotive environment, it is largely dictated by what space and mounting options are available.

"5.1 surround" means that there are five speakers that cover the main range of frequencies and one subwoofer that covers the very deep notes. Similarly "7.1 surround" adds an additional two channels to the 5.1 arrangement. In a 5.1 setup the channels are referred to as: center, front right, front left, surround right, and surround left.

The center channel is used for voice and dialogue. In certain situations where there is no room for a center speaker, a phantom center speaker can be created by feeding the output from the center speaker in equal proportions to the left and right front speakers. This results in a 4.1 setup.

To set up your car PC from the Windows XP control panel, go to: Start > Control Panel > Sounds and Audio Devices. This will bring up the Sounds and Audio Devices Properties window. You should now click on the Volume tab. Next click on the Advanced button. This brings up the Advanced Audio Properties window. Here you can select the speaker configuration.

For 4.1 surround sound, the "Surround Sound" speakers option should be selected from the drop-down menu. For 5.1 surround sound, the "5.1 Surround Sound" option should be selected. Finally "7.1 Surround Sound" should be selected if you wish to use 7.1 surround. These options are illustrated in Figures 9-8 through 9-10. These figures also offer a guide to speaker positioning.

To obtain the best surround effect in a home cinema environment, strict rules apply to speaker positioning. Unfortunately, there is

Figure 9-8 *4.1 surround sound speakers setup in Windows (Courtesy Microsoft)*

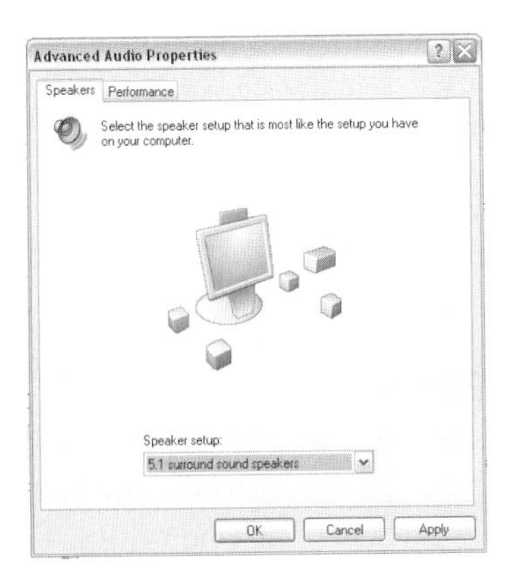

Figure 9-9 *5.1 surround sound speakers setup in Windows (Courtesy Microsoft)*

Figure 9-10 *7.1 surround sound speakers setup in Windows (Courtesy Microsoft)*

restricted space in a car audio environment and therefore compromises must be made. The following offers a guide as to how your car speakers should be positioned.

The center speaker should be positioned centrally, either mounted in the top dashboard rail or in the center console. If possible, it should be mounted as close to the video screen as possible. This is so that when actors are talking on screen, the dialogue appears to come from them rather than from the car environment.

The front left and right speakers should be positioned in front of the listener. In a car environment, the best compromise is to mount them in the front-door builds. However, if room at the sides of the dashboard is available, this is also acceptable.

The surround speakers ideally should be positioned at ear height, to the left and right of the listener. In a car environment, it is highly unlikely that this can be accommodated within the constraints of the B-pillar. Therefore the next best option would be to mount them in the rear-door builds.

The subwoofer will be larger than the other speakers and therefore requires more space. As the ear is less sensitive to the directionality of low-frequency sound, positioning of the subwoofer is less crucial. The trunk provides ample room for mounting a subwoofer.

7.1 Sound Positioning

In 7.1 surround, two additional speakers need to be considered: rear right and rear left. In a car, the best positioning of these two speakers appears to be far back on the parcel shelf at the rear of the vehicle.

All this speaker positioning is optimized for the front-seat passengers' listening experience. If the focus is to be for rear-seat passengers, then this audio scheme may need to be revised.

In terms of car audio wiring, look to Figure 9-11 for guidance on how to wire a 5.1 system with a dedicated amplifier.

In-Car Audio Troubleshooting

If you suffer from a constant whine in your car audio speakers, your system is improperly grounded. The problem is that a ground loop has been created. This is where current flows from two points that nominally have the same

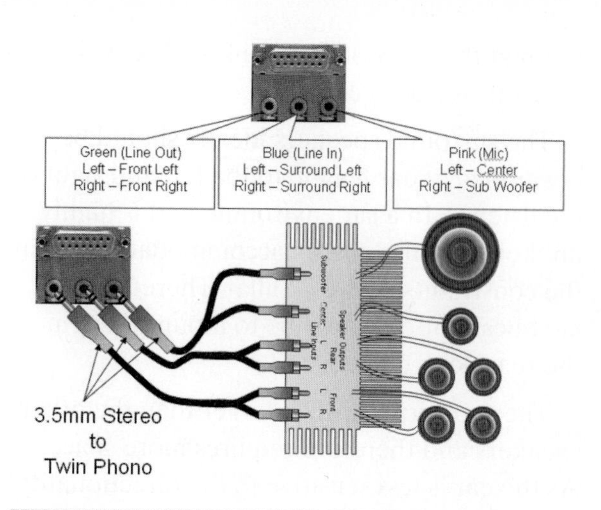

Figure 9-11 *5.1 motherboard surround through an amplifier*

potential. The reason for this is, somewhere in your vehicle, there is a resistance between the item connected to ground and the ground of the vehicle itself.

Why does this happen?

Your car PCs audio output stage is referenced to ground. This means that the ground reference for the whole audio system will be determined by the potential of ground that is connected to your head unit.

All the conductors in your vehicle have some resistance, however negligible. This includes the vehicle's body, where this is used as a ground conductor.

Wherever we have resistance, we create a voltage drop. If you cannot understand this concept, find a basic physics or electronics book.

Accessories connected to various parts of the ground-return path (that is, anywhere on the vehicle) will set up different voltage potentials over the vehicle due to the resistance in the bodywork.

As a result of the current flow through the ground loop, a DC voltage is presented to the amplifier, which amplifies it, exacerbating the effect.

In 99 percent of vehicle installations, the vehicle body itself is used as the ground-return path. The main exception to this rule is fiberglass kit cars, as fiberglass doesn't conduct electricity.

To improve the installation, a number of steps can be taken. The first, which is also the least expensive, is to use a small piece of abrasive paper to clean the metal at all the points where your vehicle audio installation connects to ground. A little rust may have formed with time, so removing any grime, and cleaning to bare metal will help form a good connection.

Remember if you remove any vehicle paint to form a good connection, you should paint over it or provide some impervious barrier once the connection is made. This will prevent the metal from rusting!

The next thing that you can do is arrange your vehicle audio wiring so it all grounds to the same point. This is a little tricky to accomplish when your audio system is distributed over the whole of the vehicle. A compromise would be to use cable to connect the points to each other, reducing any resistance in the system.

Finally, if you cannot achieve a satisfactory ground, you can purchase a ground-loop isolator. A ground-loop isolator is effectively a 1:1 wound transformer. Transformers do not allow DC to flow; they allow only AC. Furthermore, as no electrical connection exists between the input and the output section, the grounds can reference to wherever they are on the vehicle.

A ground-loop isolator is illustrated in Figure 9-12

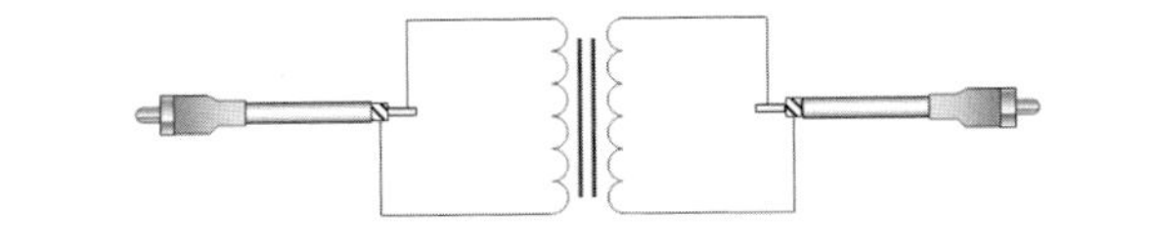

Figure 9-12 *Ground-loop isolator*

Chapter Ten

Storage Devices

Hard Disks and Car PCs

Hard disks currently provide the best dollar per megabyte for permanent, reliable storage of data. A car PC is subject to a large number of forces and vibrations, so it is imperative that the hard disk is good quality and able to withstand shock if it is to provide long-term reliable operation.

It is advisable to mount the hard drive on its side (see Figure 10-1). In the event of going over a heavy bump, the heads will oscillate up and down and not damage the disk. Were the drive mounted horizontally, a serious risk of a head-crash exists. The correct method of mounting is illustrated in Figure 10-1 and applies to both 3.5-inch and 2.5-inch hard drives.

Many constructors of car PCs have found standard 3.5-inch hard drives to be reliable in service. However in terms of space, power consumption, and shock resistance, laptop hard disks provide a better alternative. They are, however, a little more expensive.

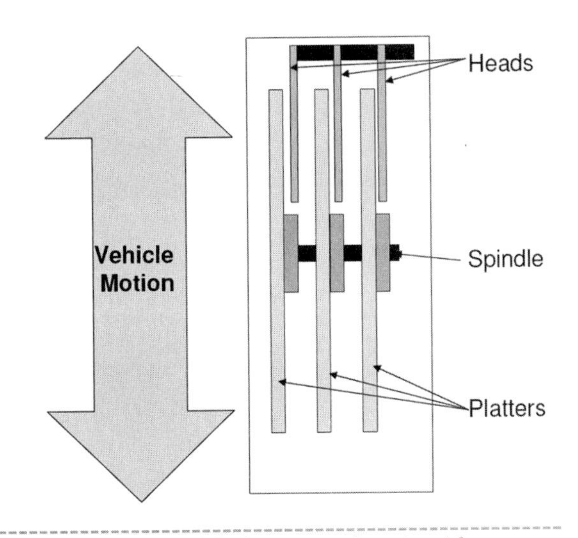

Figure 10-1 *Hard disk mounted on its side*

Laptop hard disks are available in capacities up to around 100GB; however, this figure is always increasing as the technology improves.

A good compromise between the two approaches is to use a smaller laptop hard disk (see Figure 10-2) for the operating system and all mission critical functions, and then store media on a portable external hard disk (the data is backed up on a desktop PC, therefore failure would not be catastrophic).

Interfaces

The IDE-ATA interface has served the personal computer reliably for a couple of decades. The original parallel ATA interface was introduced in the 1980s. All devices introduced since that time have been backward compatible, and as such the standard has its limitations. As drive speeds and data transfer rates have improved, the parallel *integrated drive electronics* (IDE) interface has begun to struggle with the higher data throughput. To

Figure 10-2 *Laptop hard drive (Courtesy DigitalWW.com)*

overcome the limitations of the current interface, a newer interface, the SATA, has been introduced. SATA stands for serial ATA.

Using a parallel architecture has a number of disadvantages as a result of synchronizing the signals of many wires and dealing with crosstalk and other electrical constraints. A serial bus neatly sidesteps these problems.

Another advantage in car PC setups is that serial ATA cables are much thinner and smaller than parallel cables. Parallel cables are physically very bulky and take up a lot of room preventing air circulation in confined cases. Using SATA cables avoids this problem.

If you intend to mount your hard disk remotely, another consideration should be kept in mind: SATA cables can extend up to 1 meter, whereas parallel ATA connectors are limited to 40cm.

We will be discussing car PC setups with both of these options.

IDE ATA (Parallel ATA)

The parallel ATA standard is reliable and at present used on the vast majority of hard disk storage devices. There are a number of good guidelines you can follow.

Use good quality hard disk cables. Faster hard disks require specific cables to ensure good data transfer. The better quality ribbon cables have twice as many wires joining the IDE connectors. They can be easily identified by the "finer" appearance of their ribbon cable. In these cables, every other wire is a ground. This helps to reduce interference and prevent crosstalk between data signals at high transfer rates.

You also need to note that two kinds of IDE connectors exist. Some hard drive cables come supplied with a 39-pin connector. This has a piece of plastic inserted in one of the pins to ensure that the connector will mate only one way. The other type of connector is a 40-pin connector. These are easy to insert the wrong way so *always* check before you power up. With this type of cable, it is good practice to look for pin 1 or wire 1, often identified by a red stripe along the ribbon cable and a small number molded or printed on the connector. The two types of connector can be seen in Figure 10-3.

Before selecting a hard disk, check your case and what provision it makes for hard disks.

You will also need to consider the capacity that you require to store your files. To compile a rough estimate of capacity, add megabytes and megabits for programs you will be using:

- 650 MB (minimum) per Windows XP operating system
- 700 MB (average) per GPS navigation system and maps
- 4 Mb per mp3 file (based on average song ripped at 192 k/s)
- 5,000 Mb per DVD movie (based on 2 hour movie ripped to disk without compression)

If you are going to be using a laptop style hard disk, you will need to realize that it uses a fundamentally different connection to the standard IDE drives that you are familiar with in your home PC. Figure 10-4 shows a laptop hard disk adapter.

The small female connector joins with the laptop hard disk. Be sure that the connection is rigid and that the drive is mechanically supported. It is highly inadvisable to put strain on

Figure 10-3 *Two types of IDE connectors: 39 pin and 40 pin*

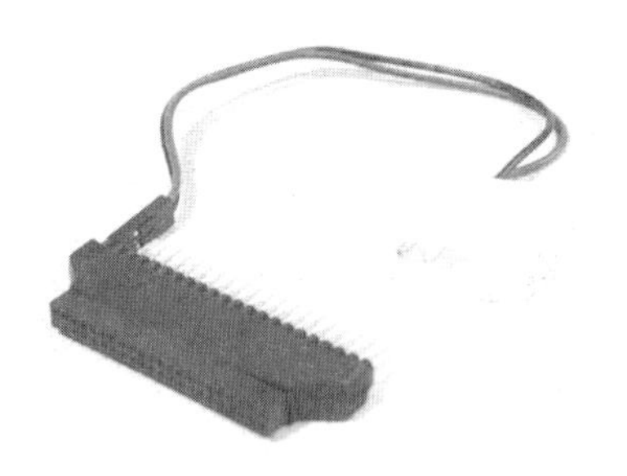

Figure 10-4 *Laptop hard drive adapter (Courtesy DigitalWW.com)*

this connection as serious damage to a laptop hard disk could occur. The 39-pin IDE connector then connects to the standard IDE cable, and the power connector connects to a standard 5.25-inch drive power connector.

Laptop hard disk drives are particularly useful for car PC applications, not only because of their diminutive size but also because a number of prefabricate cases are designed to accept them.

When choosing a laptop hard disk, you may need to consider the physical height of the disk as some cases are designed only to accept the smaller 8mm, the Travla series of cases being a good example of this.

External Hard Disks

External hard disks provide a convenient medium for storing large amounts of media and easily transporting it. With fast interfaces such as Firewire (1394) and USB 2.0 on board the MII motherboard, interfacing an external hard disk to a car PC is no problem.

Optical Storage

You probably have a large collection of CDs and DVDs. Wouldn't it be nice to be able to watch them in the car? Thankfully, standard PC drives are able to read and write to a wide range of media. When selecting an optical drive for your car PC, ask yourself what sort of media you will be using? If you only want your PC for audio, then a DVD player may be an unnecessary expense. Also think carefully about whether you *need* to be able to write to media from your car PC. You can save a lot of money by buying read-only drives.

Slot-Loading Drives vs. Tray-Loading Drives

Slot-loading drives provide an interesting alternative to the standard tray-loading drives we are familiar with. Slot-loading drives are available in both the full-size 5¼ -inch form and also in the smaller footprint for use in laptops.

In an automotive environment, a slot-loading drive may prove more durable than the fragile laptop tray-loading drives, which are prone to break. Another reason for selecting a slot-loading drive is that it more closely matches an OEM drive installation.

One of the disadvantages of slot-loading drives is that they cannot read the small 8cm media often used as promotional items and used as the storage medium for some MP3 players and digital cameras and camcorders. If you use this type of media regularly, this may figure into your decision; however, if you do not have any other 8cm devices, I would not worry about using this type of media as it is used only occasionally for e-business cards and promotional CDs.

Figure 10-5 shows the state of the art available from Digital WW, the slot-loading Panasonic DVD±R/+R DL/±RW/RAM laptop super multidrive 8X. This drive packs a lot of functionality into a very small space. It will support the following media formats:

Figure 10-5 *Slot-loading Panasonic DVD±R/+R DL/±RW/RAM laptop super multidrive 8X, from Digital WW*

- DVD RAM
- DVD+RW
- DVD-RW
- DVD+R dual layer
- DVD+R
- DVD-R
- CD RW
- CD R
- CD audio

Laptop Drives vs. Full-Size Drives

Throughout this section, there seems to be a theme of extolling the virtues of laptop hardware. The same can be said again for laptop optical drives. Once again, it can be said that they are more expensive, but by and large they also have a greater tolerance for shock, lower power consumption requirements, and a smaller package.

Laptop optical drives again use a different connector than do their full-size desktop counterparts, and an adapter will need to be purchased. The type of adapter to be used can be seen in Figure 10-6.

A proliferation of cases on the market support laptop media drives. This may be a fundamental reason for choosing a laptop optical device. However, if no specific requirement exists for a small-form-factor drive, then a

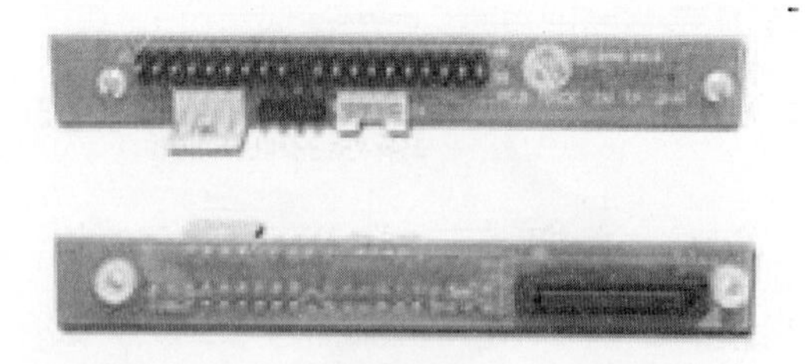

Figure 10-6 *Laptop-to-desktop IDE adaptor (Courtesy DigitalWW.com)*

5¼-inch desktop drive will more than suffice. Alternatively, it is possible to mount two laptop media drives in the space of one 5¼-inch drive. The adapter that allows you to do so (see Figure 10-7) is sold by Digital WW and makes the job very simple as well as providing an attractive black fascia.

Solid-State Storage

As memory becomes cheaper and cheaper, using solid-state memory as a storage medium for operating system and program files becomes more of a reality. In an automotive environment where a hard disk is constantly battling against vibration and bumps, solid-state storage devices with their lack of mechanical parts can prove an ideal solution.

The disadvantage of solid-state storage is that the cost per megabyte is considerably higher than for hard-disk storage.

A good compromise, where reliability is required and/or where a car PC is to be used in a harsh environment, is to use a solid-state storage device for the operating-system and system-critical files, while using a hard disk to store all media and storage-intensive files.

Figure 10-7 *Dual-mounting laptop drive adaptor (Courtesy DigitalWW.com)*

With the Mini-ITX range of motherboards, a variety of connection options exist for solid-state storage. Most obviously, the VIA MII series supports compact flash cards directly via (no pun intended) an optional expansion board. This piggybacks onto the motherboard, as can be seen in Figure 10-8.

If you are not using the MII motherboard, or the compact flash slot is already occupied with another module, do not despair! Still other ways exist to achieve solid-state bootable disks.

Probably one of the most universal methods is to boot from a USB flash key. These are available in capacities up to 1GB (at the time this book is going to print) and so can accommodate a sensible operating system installation.

Another option is to use flash memory, which connects directly to your PCs IDE socket. This can be accomplished in one of two ways:

- Use a compact flash-IDE module; insert a compact flash card in the slot and connect to the IDE socket of your motherboard as a regular device.

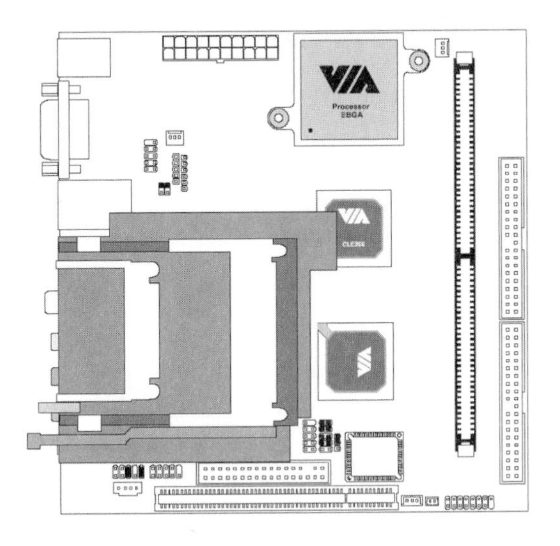

Figure 10-8 *Optional compact flash and PCMCIA module fitted to EPIA MII (Image Courtesy VIA)*

Hot-swapping ability is lost when a compact flash card is used with an adapter. It must be left connected at all times to avoid data corruption.

- The other possibility is to use a flash module that plugs directly into your IDE socket. These come supplied with all necessary interface hardware and act as a regular IDE device. They are very low profile, only slightly larger than the IDE IDC connector.

Memory Card Readers

Memory card readers offer a convenient way to exchange information between portable devices such as digital cameras, MP3 players, and the car PC system. If only occasional use of compact flash media is required, then a good compromise is to use a portable USB memory card reader. As a rule these come with about five or six interface options, are all quite common, and can be found in a variety of digital devices, such as the following:

- Compact Flash Card Microdrive
- Secure Digital Card
- Mini SD
- MultiMedia Card
- RS MMC
- Smart Media Card
- Memory Stick
- MS Duo
- MS Pro

- MS Pro Duo XD
- Magic Gate

It may sound quite obvious, but it is nevertheless worth stating: Ensure that you choose a memory card reader that will support the type of media you wish to use.

Project 44: Installing an IDE Device in Your Car PC

Warning Hard disk drives contain sensitive electronics that can be damaged by static electricity. Always take appropriate static precautions.

Integrated drive electronics (IDE) is the standard for connecting large-capacity data storage drives in most PC installations. The IDE interface is very versatile and can be used to interface hard drives and optical drives, such as CD-ROMs, DVD-ROMs, CD writers, and DVD writers, as well as other more exotic drives, like zip disks and LS-120 drives, and older technologies, such as tape drives and magneto optical drives.

In this project, we will explore the IDE interface and look at how a device must be configured to work correctly.

In recent times, the standard parallel IDE interface has become inadequate despite many years of faithful service. Although it's still quite capable for most applications, the shift is toward using SATA devices. This project is divided into two parts, as the installation methods differ. Identify the type of drive you have and follow the appropriate method.

Your drive will be a parallel IDE ATA drive if the connector on the back sports many pins.

If a flat connector is found on the rear, then the chances are your device is a SATA device. Take a peek at Figure 10-9 to aid identification. Although this illustration is based on a hard disk, the connectors will be similar for all standard *desktop* drives. For *laptop* drives, you will need to purchase an adapter as mentioned earlier.

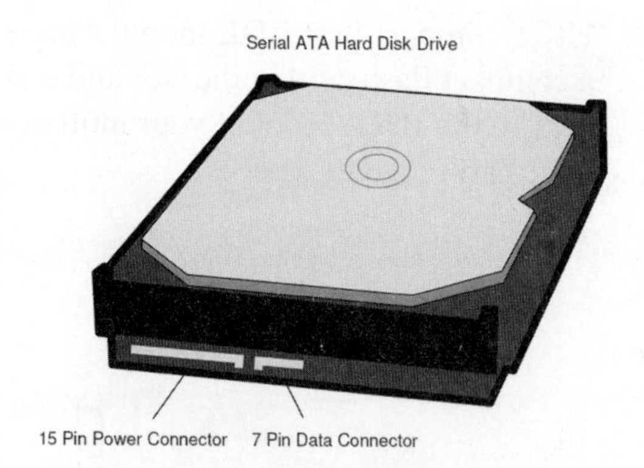

Figure 10-9 *Parallel and serial ATA connectors on hard disks*

Parallel ATA

You Will Need

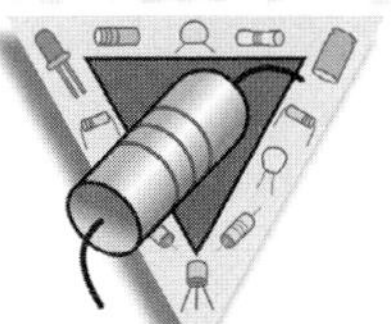

- Parallel ATA (IDE) hard disk drive
- Parallel ATA (IDE) compatible motherboard
- Grey IDE cable (80-wire for faster transfer rates)
- 6-32 UNC drive mounting screw (×4)
- Molex Y cable (optional)

Tools

- Selection of small screwdrivers

Most motherboards have two parallel ATA IDE connectors that are nearly identical no matter what model or manufacturer you choose. On the VIA MII motherboard, the IDE connectors look like those seen in Figure 10-10 and are located here.

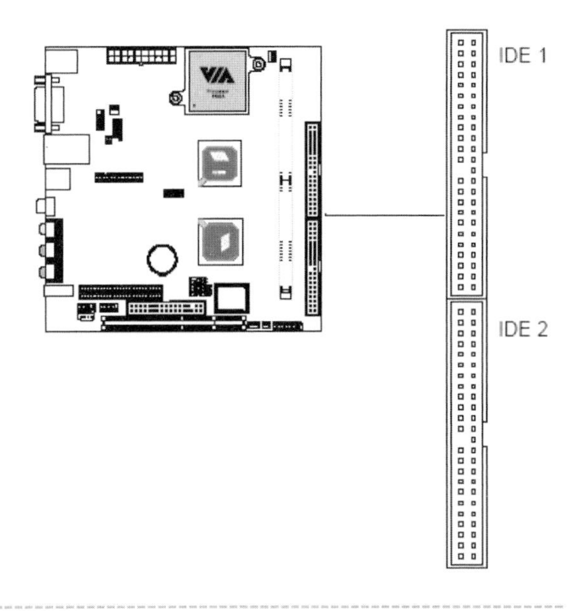

Figure 10-10 *Location of parallel ATA connectors on VIA MII (Courtesy VIA)*

As there are two connectors, we might like to say that the motherboard has two IDE channels; we will call these "primary" and "secondary." Some newer motherboards have more IDE channels, but by and large these would not be particularly useful in a car PC installation. Four channels are more than enough.

Each channel will further support two devices; these two devices can be referred to as "master" and "slave."

Master and slave devices will share the same device cable. The difference lies in the fact that jumpers, or small connections, are configured differently in order to give the device a unique identifier.

As there are two channels and two devices on each of these channels, we can surmise that a total of four devices can be supported on a motherboard of this specification.

It is not possible for two devices to share the same channel and device allocation. There can only be one primary master, one primary slave, one secondary master, and one secondary slave.

When you load your operating system, Windows will assign each of your drives a letter. Assuming that each drive has a single partition, the letters will be as follows:

- C drive = Primary Master
- D drive = Primary Slave

- E drive = Secondary Master
- F drive = Secondary Slave

As mentioned earlier, to assign the device as master or slave, the jumpers on the rear of the drive must be set. (Also one additional mode, cable select, exists in which the drive's position on the cable defines its role.)

On top of your hard disk you should find a diagram that shows you what jumpers must be set to what positions for the device to be master or slave. In the case of laptop hard disks, you will need to set the jumpers before installing the laptop hard disk adapter, as when the adapter is installed, the jumpers will be inaccessible.

The jumpers are a little sometimes difficult to move, but employing a pair of long-nose pliers will make the job significantly easier.

Once your IDE drive is configured, you now need to install it. The mechanical installation involves four screws and a screwdriver. That's about all there is to it! Just make sure the screws are not too long and that they are driven home but not too tight! Don't use an electric screwdriver; there is a serious risk of overtightening and cross-threading!

Once the mechanical installation is complete, we come to the electrical installation.

Pretty much all hard disk drives use a four-pin Molex connector for power. Connect this to a spare power connector or use a Y cable if there are no spares.

When it comes to connecting the data cable, one end goes in the motherboard. You now have a choice of two connectors. If you are using Cable Select mode, the place where you put the drive will determine whether it is master or slave. If you are not using Cable Select mode, then just be sure the data cable is plugged in the correct orientation. Once this is done, installation is complete.

Serial ATA

Installing a serial ATA drive is significantly easier than installing a parallel ATA drive. First, there are no pesky jumper settings to worry about because all drives are treated as a master device. As mentioned earlier, longer cables are also available, which is useful if you need to mount your drive separately from your car PC (for example in a removable enclosure).

Drives are hot-pluggable, which means you can even connect the drives while your car PC is running. Furthermore, connection is so easy, as the plugs are keyed to prevent incorrect entry.

Because a transition between the parallel and serial IDE standards currently exists, many new SATA drives still have the Molex four-pin power connector to retain backward compatibility.

CAUTION

When using the four-pin Molex connector to power your drive, hot-swap functionality is lost.

Because at present there are only a limited range of power supplies with support for SATA drives, the easiest way to power the drive is by connecting a four-pin Molex connector to 15-pin SATA power connector. The leads incorporate power regulation, as SATA drives require a 3.3-volt line in addition to the 5- and 12-volt feeds of the Molex connector. By using this approach, hot swapping is possible if the drive and motherboard support it.

You Will Need

You Will Need

- SATA-compatible motherboard
- SATA-compatible HDD
- SATA data cable (seven-pin plugs)
- Molex SATA power supply adapter
- 6-32 UNC drive mounting screw (×4)

Tools

- Array of small screwdrivers

Installing SATA devices is pretty easy. Remember that one device is supported per cable unlike with conventional IDE cables, which can have a master and a slave.

Check with your manufacturer's instructions regarding what length screws are required. Screws that are too long may damage the inside of your hard drive if driven fully home.

Mechanical installation is the same as for any other device. Choose your location, align the holes in the case, and screw the screws home so that they are all the way in but not too tight.

First, connect one end of the adaptor cable to the Molex-SATA power adapter cable. Next, connect the SATA data cable to the motherboard's SATA connector. Look back at Figure 10-11 if you have an EPIA SP motherboard and you want to locate the SATA connector.

On the drive end, first connect the power cable, followed by the data cable.

When it comes to software setup, you will need to ensure that the BIOS is correctly configured. Take a look at Figure 10-11.

This application is specific to the EPIA SP motherboard, check with your individual manufacturer's instructions if you are using something else.

You need to be sure that "On Chip SATA" or something similar is enabled. When it comes to SATA mode, you have a couple of options. When the option is set to IDE, you can use two parallel IDE hard disks and two SATA drives. When the option is set to RAID, you will find that only the SATA drives will support the RAID.

Your operating system may also require you to install a SATA driver. Once this is complete, installation is finished.

Troubleshooting

The serial ATA cables are not shielded, therefore a little care must be taken when operating the drives.

Make sure when you operate the SATA drive that it is inside a PC enclosure, as the interface cables are more prone to pick up

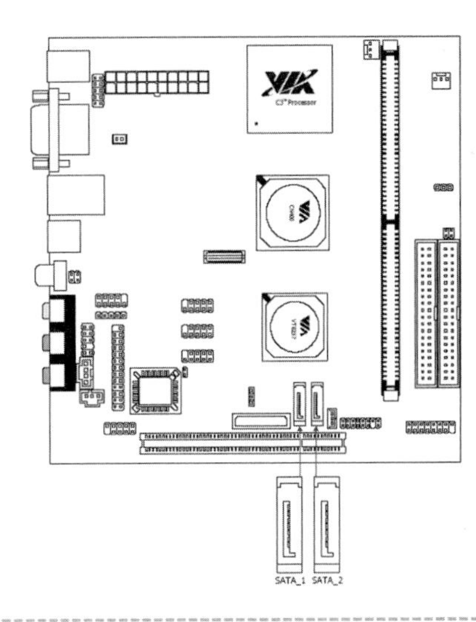

Figure 10-11 *SATA location on VIA EPIA SP (Courtesy VIA)*

electrical noise. Furthermore, don't tie-wrap SATA cables together or, for that matter, PATA cables. Try and keep your cell phone away from SATA installations, as the cables are prone to picking up electrical noise, with corrupt data being a possible result.

If your drive is not working properly, these troubleshooting tips may help solve the problem.

Check that your drive spins up. When a drive spins, you will hear the mechanism purring, followed by a few light clicks. If this does not happen, check the power and interface cables.

If the drive spins up but the system does not recognize it, check that the SATA features are enabled in the system BIOS. See Figure 10-12. If that does not solve the problem, check that the SATA drivers are installed correctly in the operating system.

A Note on Floppy Disk Drives

Floppy disk drives are now largely redundant due to their limited capacity. In a car PC system, they are of limited use. Some motherboards, especially of the small form factor type do not include a floppy disk drive connector, as they are largely unused. The VIA MII motherboard *does* have a floppy drive connector, should you require it.

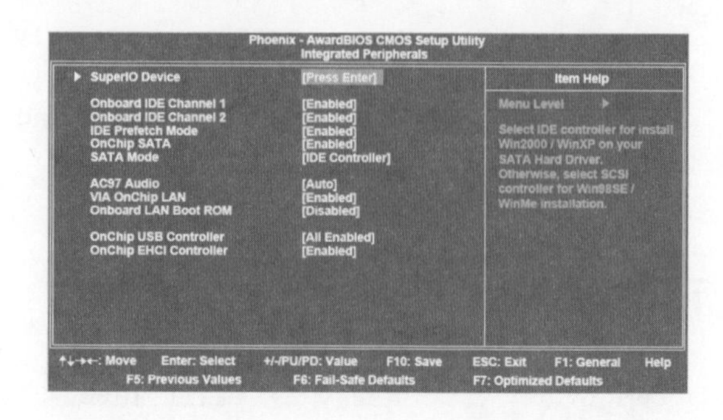

Figure 10-12 *SATA settings in BIOS screen*

Chapter Eleven

Speech Recognition

Speech recognition is the ultimate in geek chic if you can get it to work properly. A car that answers to your every utterance is the utopian dream of many a sci-fi TV series or film. However, achieving this in practice is a little more difficult than it may first appear. Speech recognition is possibly the most ideal interface method in a perfect world with perfect technology. It requires little concentration and is natural, because, after all, it is the way that we communicate with each other.

In this chapter we will explore the applications for speech recognition in your vehicle. We will look at a few software applications for your car PC and also look at circuitry that you can implement independently of your car PC to enable you to control car functions with the sound of your voice.

First, we are going to look at some of the technical aspects of speech recognition.

Speaker-Dependant or Speaker-Independent Speech Recognition

Speech recognition falls broadly into two categories:

- Speaker-dependant recognition
- Speaker-independent recognition

You do not have to be the sharpest knife in the drawer to deduce that a speaker-dependant system is dependent upon the particular speaker who has trained the system, whereas the speaker-independent system should function regardless of who the speaker is.

To broadly illustrate this, imagine dictating a letter to your home PC. The chances are that you have trained the software package by reading a number of short extracts to allow it to adjust to the sound of your voice. This is an example of a *speaker-dependant system.*

Now imagine a voice-activated voicemail machine that requires you to say a number after the tone. The speech recognition must be able to recognize a wide variety of voices. This is an example of a *speaker-independent system.*

Hardware-Based and Software-Based Speech Recognition Systems

Within these two types of systems, we could further classify another distinction:

- Hardware-based solutions
- Software-based solutions

Hardware-based solutions have a little less flexibility than do software-based solutions and can often recognize a much more limited set of commands. Often they are based around an *application-specific integrated circuit* (ASIC). For applications that are relatively undemanding, hardware-based solutions can remove much of the complexity that a computer would entail.

Software-based solutions are programs that run in the back of the *graphical user interface* (GUI) environment unnoticed; however, they do occupy a quite significant portion of memory and are processor intensive. A common implementation of speech recognition software is to use a word key to activate the software. This could be anything from Aardvark to Zebra. The computer constantly listens for this word. Upon hearing the word, the computer will automatically switch on speech recognition and interpret all of the data that follows the word.

How the speech is recognized also forms an integral part of the classification of a speech recognition system.

The styles of speech recognition include the following:

- Isolated speech
- Connected speech
- Continuous speech

Isolated

In this style of speech recognition, only individual words are recognized. This is the main type of recognition you will work with in the hardware-based speech recognition project in this chapter.

Connected

In connected speech recognition, a number of words in short succession can be recognized. Using the longer word length, it is possible to experiment with this type of speech recognition with the hardware-based project in this chapter.

Continuous

Continuous speech recognition systems can recognize naturally spoken, continuous, unbroken speech. Some PC-based software applications are approaching reliability with continuously spoken speech; however, further development is still required.

In this chapter, we are going to look at a couple of different ways of tackling speech recognition in an automotive environment. The first will focus on building a hardware solution that can be used to control simple devices in the car environment. The circuit has much flexibility as it is used to actuate 12-volt relays. This means that it is far reaching in scope and potential.

A number of possible applications exist:

- Controlling interior lamps
- Controlling electric windows
- Controlling electric seats
- Switching amplifiers, screens, and other AV equipment on and off
- Powering up a car PC on a given command
- Acting as an alternative indicator lever (i.e., "say left/right" before the junction)
- Switching on the heater and air conditioner

PC-Based Speech Recognition Systems

Let's talk about some of the functionality that Windows XP has built in that allows users of car PCs to perform speech recognition functions.

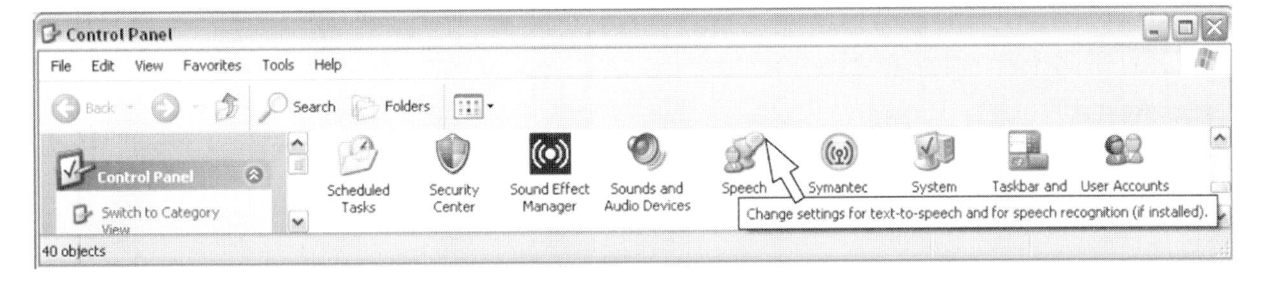

Figure 11-1 *Control panel, speech settings (Courtesy Microsoft)*

First, take a little browse in the control panel. You will see the speech option (see Figure 11-1). Clicking on this should give you a few further clues as to the functionality built into Windows.

From the Speech Properties window that appears, you will be able to find and change the program used for speech recognition by your car PC, and you will be able to train and edit speech profiles. This page is also useful for setting up your microphone correctly to get a clear signal with the correct amount of gain.

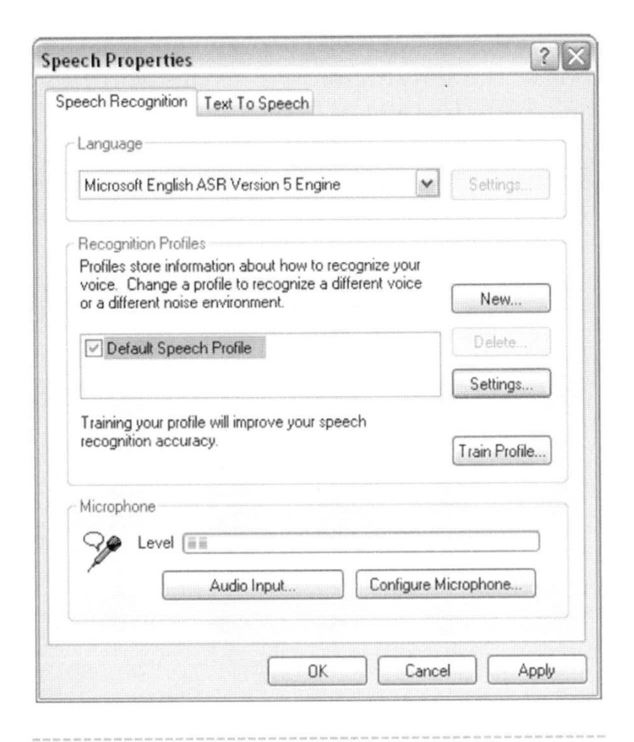

Figure 11-2 *Speech Properties window (Courtesy Microsoft)*

Windows XP supports a variety of speech recognition engines. There are a lot of packages on the market that are designed to work with a standard windows interface and will allow you some capability to control your car PC. If you search on the mp3Car.com forums, one of the users has created a program called

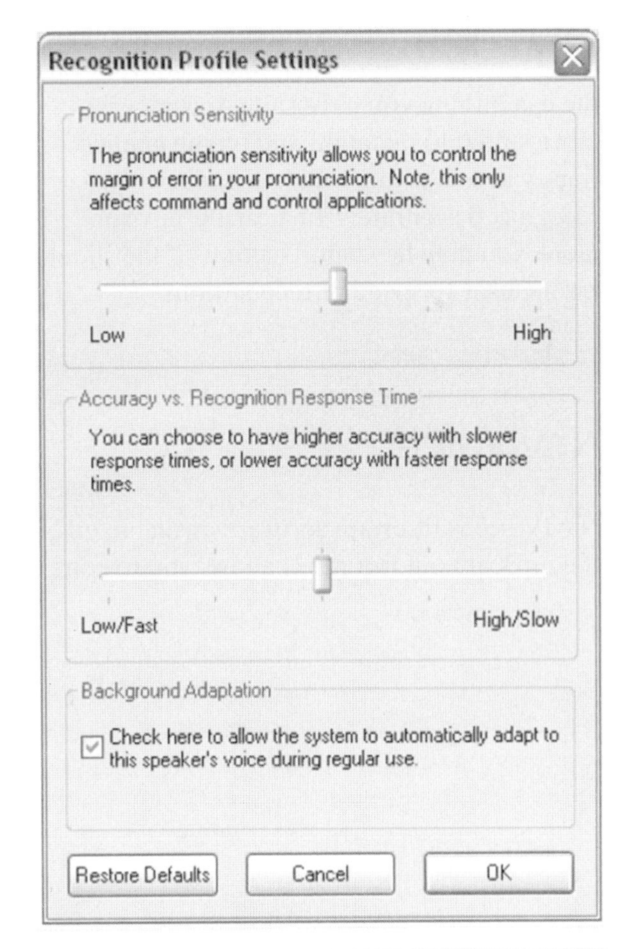

Figure 11-3 *Recognition Profile Settings (Courtesy Microsoft)*

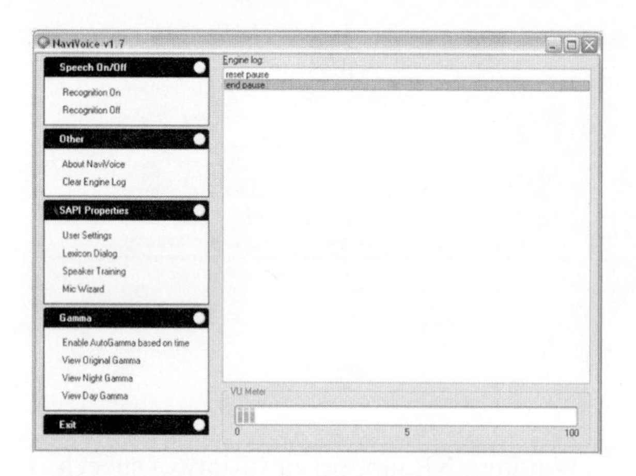

Figure 11-4 *NaviVoice screenshot (Courtesy mp3car.com)*

NaviVoice, which is tailored for use with the car PC.

You may find that your PC takes a little while to respond to certain commands; this is because the speech recognition software takes some time to analyze your voice and work out the exact thing you have said. Accuracy is more suited to dictation and typing applications where it is imperative that each word is recognized accurately. In a car PC environment, you may have more success if the slider is set closer to speed of recognition.

NaviVoice

NaviVoice is freeware software available on the mp3car.com forums. It allows you to con-

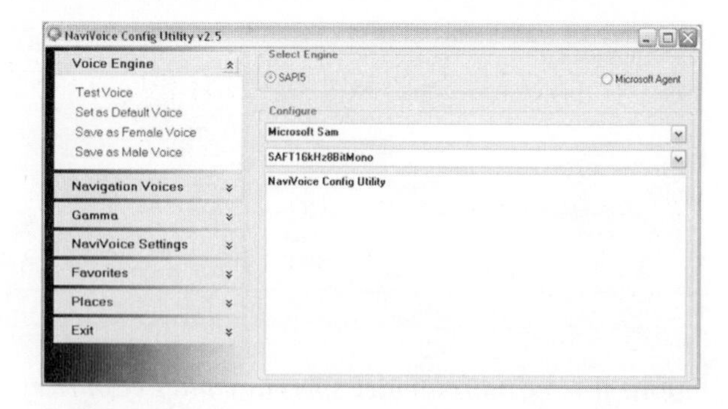

Figure 11-5 *NaviVoice Config Utility (Courtesy mp3car.com)*

trol common car PC applications with the sound of your voice. Some of the presently supported applications include the following:

- Routis
- IGuidance
- Frodoplayer
- Mapmonkey
- Road Runner

To use the software, you simply speak a keyword, which is user-programmable. As soon as the software recognizes this word, it puts it into listening mode.

Project 45: Building a Speech Recognition Module

This speech recognition project is based on the HM2007 IC, which is an ASIC. It contains an onboard CPU and is capable of speech recognition. It boasts many features.

It is a single-chip voice recognition system requiring only external memory and a minimum of components. The chip is speaker dependent. However, a clever trick can be used to enable this circuit to be speaker independent, as we shall see later on.

Warning Please Note: This circuit can only really be recommended for use with "non-mission-critical" applications. If you were to say "lights," and the interior lights were not to come on, it would be no great catastrophe. However if you were to say "brakes," and the brakes did not react, you would be faced with a different situation entirely.

The circuit comes supplied in kit form by Images SI Inc. (see Appendix A for contact information). Images SI Inc. incidentally also makes a nice relay interface board (see Figure 11-7), which makes interfacing to the car a breeze.

The board supplied from Images SI Inc. is really a development board and allows you to experiment with the sophisticated HM2007 chip.

The chip will recognize 40 words with a word length of 0.96 seconds. In this mode, we could reason that the speech recognition board was working in the style of isolated speech recognition. Alternatively, the circuit can be set up to recognise twenty 1.92-second words of speech. As it is possible to feature a few short words in this time frame, the speech recognition circuit could be said to be working in the connected style of speech recognition.

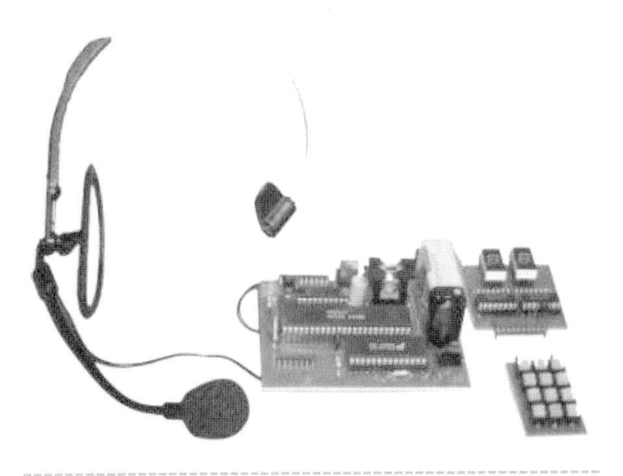

Figure 11-6 *SR-07 speech recognition kit (Courtesy Images SI Inc.)*

The HM2007 supports two modes of operation, manual and CPU connected. In this project we will be looking at the chip in its manual mode of operation; however, nothing says you can't use it separately in the CPU-connected mode with a suitable microcontroller. The manufacturer's datasheet will be an invaluable resource to those wishing to experiment and can also be obtained from Images SI Inc.

The circuit uses the HM2007 chip with a separate *static random access memory* (SRAM) chip used to hold processing data. The chip provides for two *binary coded decimal* (BCD) outputs in the manual mode. These outputs can be connected to a decoder and pair of seven-segment displays (supplied) for debugging purposes or directly to logic for further processing. The relay board supplied by Images SI Inc. (www.imagesco.com) connects to these BCD outputs and allows the user to control up to 10 devices.

Figure 11-7 *SRI-02 speech recognition interface kit (Courtesy Images SI Inc.)*

Circuit Operation

To operate the circuit, you will use the keypad to input data and the LED display to view the output. Later when you are satisfied with your results, you can connect the relay boards to the BCD outputs to control real devices.

Upon powering up the circuit, the LED display should read 00. This indicates that the circuit is ready to go. The red LED should also be lit at this point.

You need to train the circuit to recognize different words. To do this, input a number between 1 and 40 followed by the hash key.

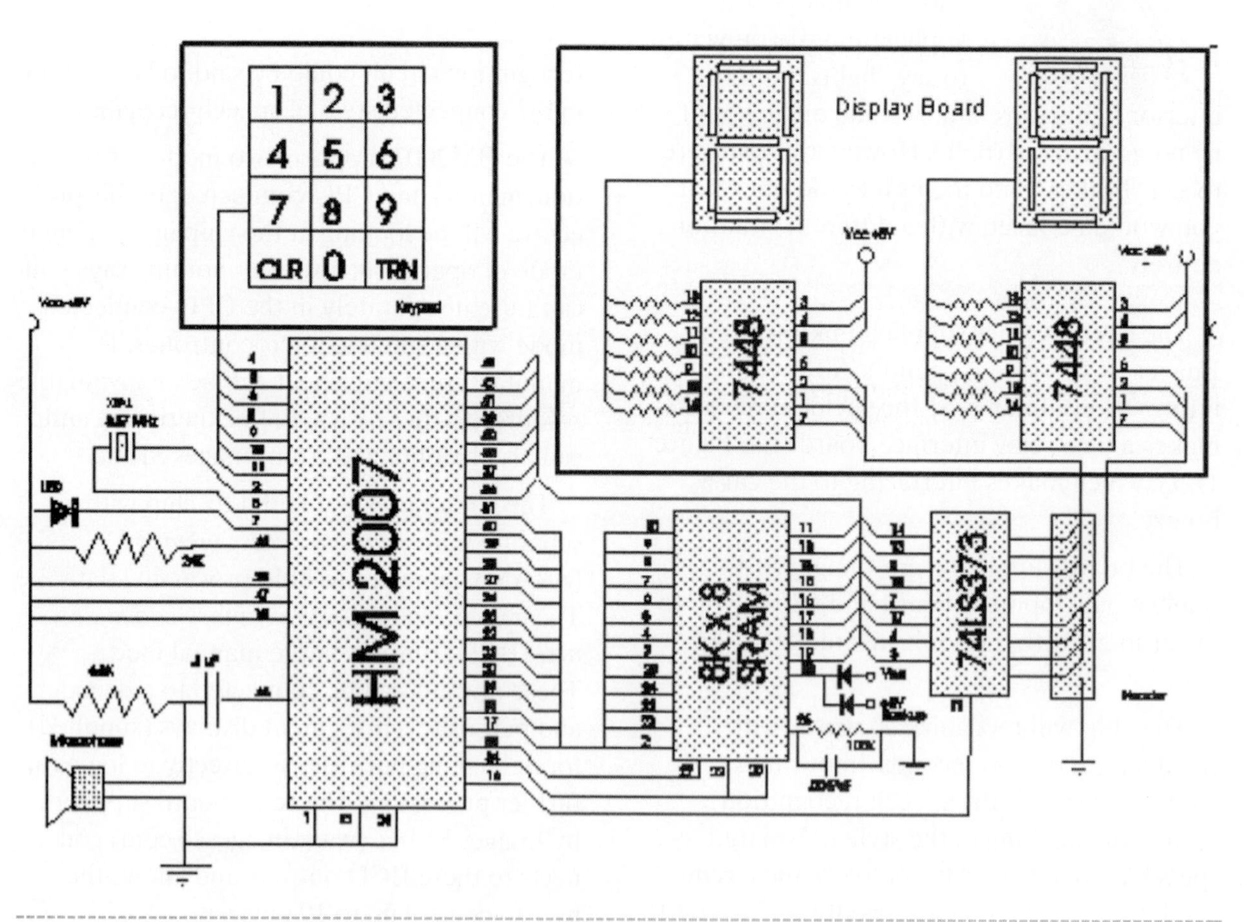

Figure 11-8 *Speech recognition circuit diagram (Courtesy Images SI Inc.)*

The red LED should blink on, indicating the circuit is ready to receive the word. Think of it, if you like, as a recording light. Speak clearly into the microphone in your normal tone of voice.

Just because there are 40 word spaces, it doesn't mean you have to use all of them. It is quite acceptable to use only the amount of word spaces required for your application.

When you have finished training the circuit, you are ready for it to recognize your spoken command. Simply speak into the microphone. When you do this, the number of the word that you programmed should come up on the LED display. For example, if you programmed "radio" into word space 1, 01 will be displayed when you say "radio." The signals presented at the two BCD outputs will be 0000 and 0001 representing the characters 01. This can be decoded by logic and used for your own control applications.

Speaker-Independent Recognition

By training a number of word spaces to one word using different individual's voices, a primal form of speaker-independent recognition can be achieved. Using our example of the word "radio," the system could be trained to occupy spaces 01, 11, 21, and 31. If you are using this setup, you need to decode only one of the BCD outputs. The one exception is error codes (see troubleshooting), which would need to be removed from the final output using logic.

Troubleshooting

The chip has a number of error codes reserved. There are 55, which are displayed when a word is too long. Conversely, 66 is displayed when the word is too short, and 77 is displayed when no match can be made to an existing word.

Chapter Twelve

GPS

What Is GPS?

GPS, or *global positioning system*, allows any individual in the possession of a GPS receiver to locate his or her position on Earth with incredible accuracy. GPS can also provide information on altitude as well as longitude and latitude.

The global positioning system was launched by the U.S. military as a defense navigation system. It is a constellation of 27 satellites that constantly orbit the earth. Of these 27 satellites, 24 are in active use, while the other three provide a backup in case of failure. The system is now in the public domain and available for everyone to use.

To use the system, there are no fees or licenses required; anyone with a GPS receiver can take advantage of incredibly accurate global positioning. GPS has been employed in OEM car navigation systems for a number of years; however, until recently the technology has been beyond the domain of the home hobbyist. In the past couple of years, an explosion has happened in the proliferation of GPS devices, and the technology is now mature and accessible to a wide range of users.

How Does GPS Work?

To calculate your position, a process called *trilateration* is employed.

On each of the GPS satellites are accurate atomic clocks. These are constantly transmitting the time to Earth along with a unique identifier called *a pseudorandom code*. The receiver has a quartz clock, which is kept accurate by calculating the average of the times transmitted by all the other satellites.

Radio waves travel at the speed of light. By analyzing the discrepancy between the time transmitted and the actual time of the onboard clock, it is possible to calculate how far away the satellite is.

As we know, we are *x* distance away from *y* satellite; we could be at any point on a sphere surrounding the satellite. And because there are many satellites in the sky, we can expect to be able to receive a signal from one or more satellite.

Imagine a sphere surrounding each satellite. This sphere represents a signal that has taken a fixed amount of time to travel from the satellite. The point at which the spheres intersect defines a point in 3D space. The fact that the earth is a sphere gives us another frame of reference. By looking at the possible intersection points of the spheres and then calculating which points are realistic as they lie on the earth, we can calculate our position (see Figure 12-1).

Which GPS Device?

For most people a USB mouse will be the cheapest and most efficient solution. A USB

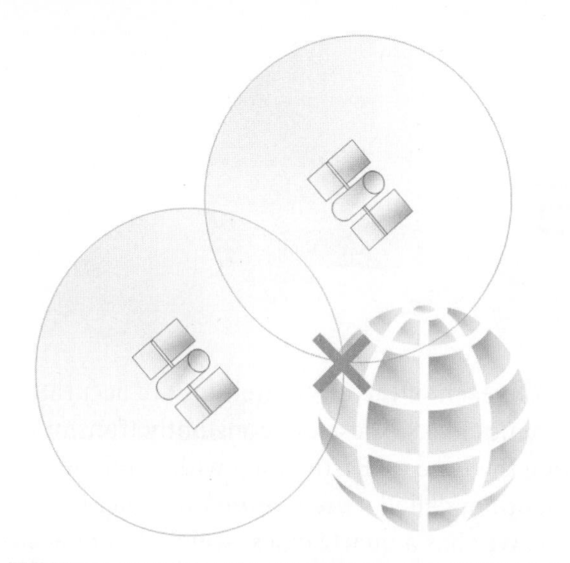

Figure 12-1 *Explanation of how GPS pinpoints position*

Figure 12-2 *Deluo Compact Flash GPS Universal (Courtesy Deluo)*

GPS mouse has no display or controls, just a USB or serial lead to connect to your PC. The aerial is generally integrated as part of the device. These aerials are small and unobtrusive and can be mounted using either a sucker cup or magnetic base.

It is important to bear in mind that serial ports are being phased out on newer motherboards; USB is heralded as the way forward for interfacing peripherals. This means that connecting serial devices to newer PCs will become harder in years to come. USB devices are not much more expensive than serial devices anyway, so the extra investment will prove worthwhile.

In addition to this, we are seeing a greater range of products on the market as a result of PDA-based GPS software. Compact flash GPS modules such as the Deluo Compact Flash Universal (illustrated in Figure 12-2) will interface directly with the EPIA MII series of motherboards, making use of the compact flash slot.

An alternative approach uses the wireless interface Bluetooth (discussed further in Chapter 12 on wireless connections) to connect your GPS device to your car PC without wires. The beauty of using a wireless interface is that it can be carried around on your person and used with many devices, such as on your PDA and mobile phone, as well as on your car PC.

Again Deluo comes to the rescue with the superb Deluo Bluetooth GPS Lite (shown in Figure 12-3).

What GPS Software?

For a run-down of GPS software, see Chapter 15, "Car PC Software." A vast array of commercially available software mapping programs now exists: Microsoft MapPoint, Microsoft Streets and Trips, AutoRoute, Destinator, iGuidance, Routis, Delorme Street Atlas, and CoPilot. We will be looking at these a little later in this chapter.

Figure 12-3 *Deluo Bluetooth GPS Lite (Courtesy Deluo)*

Project 46: Installing a GPS Mouse

You Will Need

You Will Need

- USB GPS mouse
- USB GPS mouse driver software

This guide has been written specifically for GPS mice that use the inescapable BU-303 SIRF chipset. However, devices that use other chipsets should have a very similar installation procedure.

When installing the windows driver for the BU-303 GPS, you are actually installing a driver for the USB to serial chipset. The driver's function is to take the information received from the USB port and map it to one of the computers COM (or serial) ports. This gives the illusion that the information is actually coming from that port, which makes configuration of the software much easier.

The device installs much the same as any other USB devices. A variety of methods can be used, but by and large the driver installation procedure will involve inserting a CD with an autorun program or looking for the unknown device in the control panel and then updating the driver.

Whichever procedure you follow, you must note which COM port you map the device to.

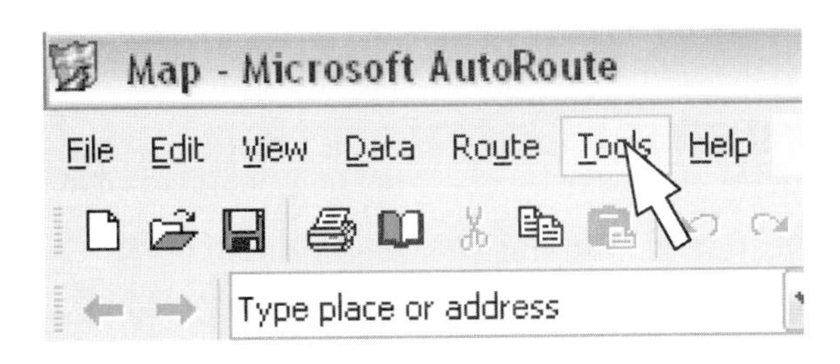

Figure 12.4 *GPS configuration, AutoRoute, step 1 (Courtesy Microsoft)*

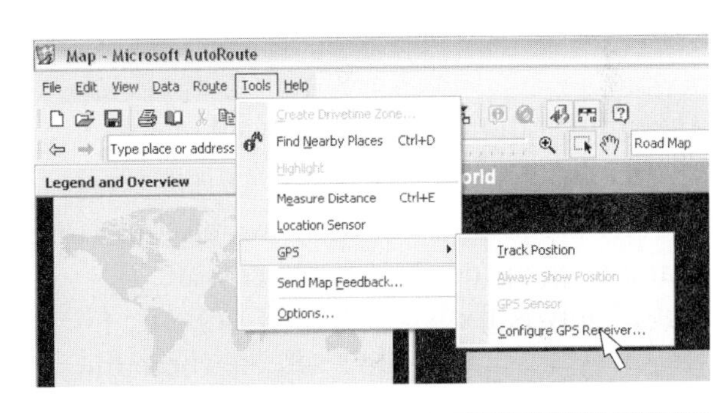

Figure 12.5 *GPS configuration, AutoRoute, step 2 (Courtesy Microsoft)*

Figure 12.6 *GPS configuration, AutoRoute, step 3 (Courtesy Microsoft)*

This information should be placed in a secure place.

Software procedures differ widely, but as an example, the procedure for Microsoft AutoRoute is Tools > GPS > Configure GPS Receiver. This brings up a screen where the number of the COM port being emulated by the USB driver should now be input.

Are there any disadvantages to a home-made satellite navigation system compared to a commercially available one? Yes. Nevertheless, many hobbyists have built satellite navigation systems based on PCs and reported that they are very usable and provide good accuracy.

Unfortunately, a few technical differences exist that will leave hobby systems a few steps behind commercial OEM systems. Commercially built systems derive position information not only from a GPS signal, but also from inertial sensors, such as accelerometers and gyroscopes, as well as from the vehicle's odometer feed. This raises a few minor problems.

GPS location may be slightly different from actual location; this can largely be corrected in software. The lock-to-road function in many software packages changes the GPS position to the nearest road location by calculating the nearest logical position according to the map. This prevents the car from being displayed off road. This is one of the software algorithms that helps make a homebuilt GPS navigation system match the performance of a commercial one. The commercially fitted OEM option, however, does have additional information, and therefore holds the edge in terms of performance and accuracy.

Another small problem arises in tunnels or other areas where GPS reception is poor or lost. Without the additional information provided by inertial sensors or the odometer, the car has no way of knowing where it is. This is only a problem for the duration of the tunnel; a signal lock should be regained quickly when the obstacle has passed.

At the time this book is going to press, a single GPS device remains the best way to interface to a car PC; this is because commercially available software does not presently exist that will support other inputs. However, the hardware is available to read the vehicle's odometer output and gather inertial data. The following are links to Web sites of companies that have developed solutions to this problem. They are ideal as a starting point for research and development:

- www.whitebream.com/p801.shtml
- www.pointresearch.com/vnu_sheet.htm
- www.trimble.com/placer450455.html
- www.u-blox.com/products/sbekit.html

Internal Construction

Despite its diminutive size, the small GPS mouse is a powerhouse of technology. It has a number of components:

- The antenna picks up the signals broadcast by the GPS satellites.
- The downconvertor converts the radio frequency signal into an intermediate frequency signal, which is fed into a correlator, which processes the information received. This is then fed into a microprocessor, which is part of the chipset, to convert the information into an intelligible form that our car PC can understand.
- In the case of the USB mouse, an additional serial-to-USB chip is employed to convert the serial output of the SIRF chipset to a USB output.

Selective Availability

GPS is the result of a massive investment from the U.S. armed forces and was originally designed with military application in mind. To this end the signal was degraded intentionally by inserting random erroneous information in the transmissions to prevent the system from being used by the enemy. This practice was discontinued in May 2000 and allows the GPS system to function with much greater resolution and accuracy.

Onboard Diagnostics

Many thanks go to Matt Williams and Mike Fahrion at B&B Electronics for their invaluable help and assistance; without them this chapter would not have been possible.

Introduction to OBD

Onboard diagnostics were originally introduced to cars as a measure to control emissions, but since OBD's inception, it has rapidly expanded to encompass all of the car's sensors and diagnostic inputs. Most cars manufactured from the mid-1990s onward will have OBD systems.

Three standards are in common use:

- ISO 9141
- SAE J1850 VPW
- SAE J1850 PWM

All cars that have adopted the OBD-II standard have a socket or receptacle that is accessible from the driver's seat. The location of this will be shown in your car owner's manual. If this information is not present, a friendly mechanic or dealer should be able to tell you. The OBD connector interfaces to a car PC via a plug-in interface lead.

Project 47: Unlocking the Secrets of Your Automobile's Central Nervous System

Warning

Do not attempt to operate a computer or analyze readings while driving. Get an assistant to help, or data log the data, and then you can analyze afterward. Many readings can be taken while the vehicle is stationary.

Introduction

Have you ever wondered what causes the check-engine light to turn on, whether the problem requires immediate attention, or what must be done to turn the light off? Are you concerned about passing your emissions test the first time? As automobiles have become more sophisticated, answering these types of questions has become increasingly difficult without the right tools. AutoTap® was designed to demystify the complexity of your automobile and empower owners to quickly diagnose issues accurately and inexpensively. AutoTap makes the dealer's diagnostic data available to automobile owners and thereby gives owners the power to monitor, diagnose, and repair today's vehicles.

You Will Need

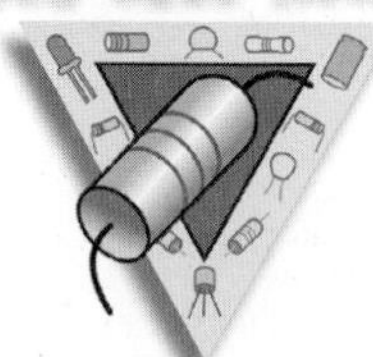

AutoTap OBD-II diagnostic scantool kit (Figure 13-1)

Connector cable

Registration key

Software

PC running Windows 98, ME, 2000, or XP with 50 MB of free hard drive space and 32 MB of RAM OR a Palm PDA running O/S 3.5 or higher and 8 MB of RAM

1996 or newer car, light truck, or SUV that is OBD-II compliant

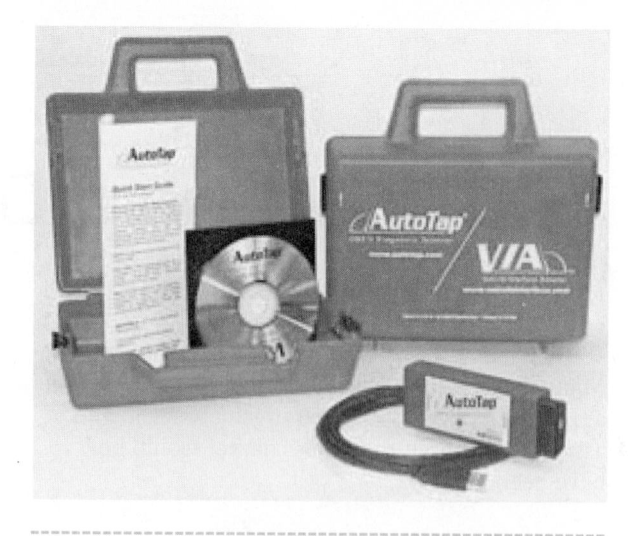

Figure 13-1 *AutoTap OBD-II diagnostic scantool kit (Courtesy AutoTap)*

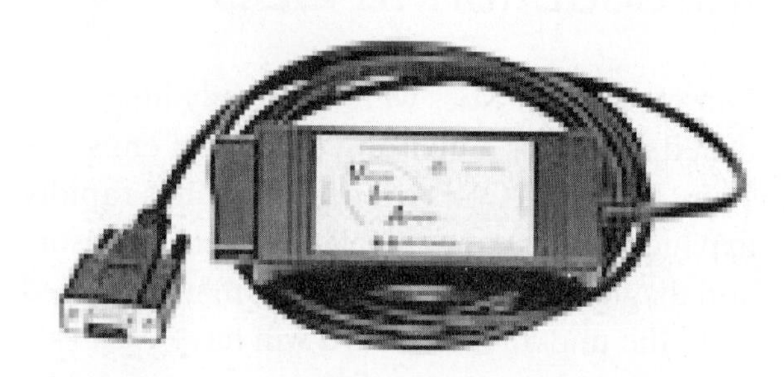

Figure 13-2 *AutoTap cable (Courtesy AutoTap)*

Figure 13-3 *OBD-II port (Courtesy AutoTap)*

Tools

No special tools are required.

All cars and light trucks built and sold in the United States after January 1, 1996 are required to be OBD-II equipped. If your vehicle is OBD-II compliant, you will find a note on a sticker or nameplate under the hood stating "OBD-II compliant."

Getting Started

To get started, you will have to install the AutoTap software onto your PC or PDA.

Installation is easy, just insert the AutoTap CD, and the software will automatically install. Once the software is installed, simply plug the AutoTap cable (see Figure 13-2) into the OBD-II port (see Figure 13-3) located low on the dashboard on the driver's side. You will need to turn the vehicle's ignition to the run position and double-click on the AutoTap icon to initiate the registration process. During your initial startup, you will be prompted to enter the registration code. After your registration code has been accepted, AutoTap will continue the registration process by automatically reading your *vehicle identification number* (VIN).

Now you are ready to demystify the sophisticated electronics in your automobile. Start your car and launch the AutoTap software. Once it queries your vehicle for supported parameters, you can start to play. You'll find hundreds of parameters—one for virtually every sensor under the hood.

Creating Your Personalized Dashboard

Each parameter can be displayed in a graph, gauge, or table. Spend some time building your ideal digital dashboard. To add a gauge, either right-click on the main screen or select Display on the menu bar, and then select Add Gauge. Your next step is to choose the parameter(s) you want to monitor, which can be accomplished by selecting Change and then highlighting the parameter to monitor. In the following illustrations (Figures 13-4 and 13-5), sample gauges for engine speed, vehicle speed, coolant temperature, and calculated engine load have been created. To modify range, style, and color, simply click Advanced and then follow the onscreen directions. By following the same process, you can quickly create tables or

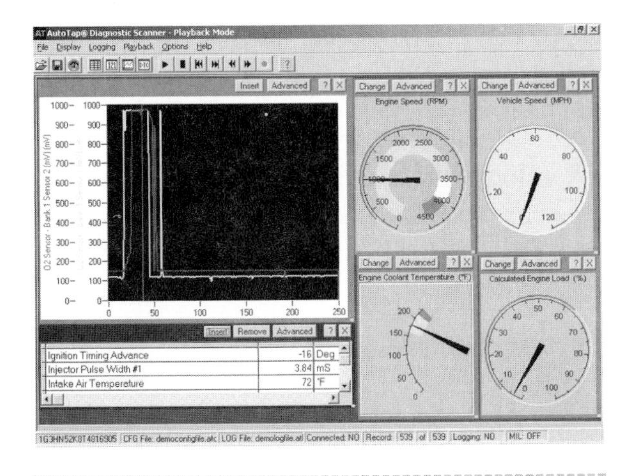

Figure 13-4 *Sample gauge panel I (Courtesy AutoTap)*

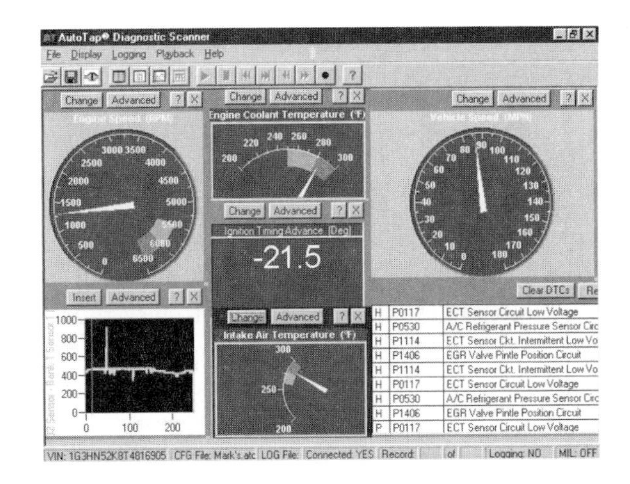

Figure 13-5 *Sample gauge panel II (Courtesy AutoTap)*

graphs to monitor or record the parameters important to you.

Hundreds of useful applications exist for AutoTap. I have listed a few to stimulate your creativity.

Table 13.1

Everyday applications for the "shade-tree mechanic"

Application	*Parameter*
Towing a trailer	Monitor transmission temperature to prevent premature failure
High performance	Monitor knock retard and add higher-octane fuel as required
Mega audio	Monitor battery voltage to keep tabs on your alternator
Vehicle repair	Monitor *diagnostic trouble codes* (DTCs) to identify and repair failed components or sensors
Emissions readiness	Monitor the check-engine light and read DTCs
Check-engine light	Monitor the *malfunction indicator lamp* (MIL) status

Beginning in 2002 a number of states announced plans to change their emissions testing programs to OBD-II. Instead of doing a tailpipe emissions check on a dynamometer, an OBD-II check is a simple plug-in test that takes only seconds and is more accurate. It is easy to determine if your automobile is ready to pass an OBD-II emissions test.

Emissions readiness and the check engine light are related. If your check engine light comes on in your OBD-II automobile, you have an emissions-related failure. With AutoTap you can quickly determine the status of the check engine light. At the bottom of the screen, the *malfunction-indicator light* (MIL) will be either on or off based on whether the check engine light is on or off. Better yet, you will be provided a list of *diagnostic trouble codes* (DTCs) with descriptions of the reasons the MIL might be on. To turn the light off, just select Clear DTCs, which will be located at the top of the table that summarizes the DTCs. If the faults persist, you will be able to use the DTC descriptions to guide you through a quick and accurate diagnostic and repair.

Troubleshooting

AutoTap has been on the market for many years and is used every day by thousands of loyal customers. If you encounter trouble getting up and running, validate that the software was installed correctly, the vehicle registration process was followed, and the USB drivers have been loaded.

If the software does not load automatically, click the Start button, click Run, type D:Setup, select Install Software, and follow the on-screen instructions. To avoid registration problems, please make sure that you connect the cable between your computer and vehicle and have the ignition in the Run position when you enter the registration code.

After the registration code has been accepted, AutoTap will attempt to read your vehicle's VIN automatically. You may also select Connect Without Using VIN. If you select this option, only generic parameters will be available. If you complete the registration and connection process correctly, you will get a message at the bottom of the screen that says Connected: YES.

Sometimes customers have older computers and they will get messages stating that AutoTap was unable to connect to the vehicle; it might say Connected: NO. If you followed the registration and connection process outlined here, chances are that you need to load the USB drivers. USB driver software is included in the CD for the AutoTap software. To load the drivers, find the file ATUSB_Drivers.zip on the AutoTap CD and double-click on it.

If a window titled Lost Communication pops up, check all of the connections between your car PC and your car. A connection may be loose.

If you have any further troubleshooting problems, email Autotap at support@autotap.com.

Diagnostic Trouble Codes Explained

It is useful to understand how a DTC is composed and what all of the sections mean to help you better understand your vehicle (see Figure 13-6).

The first character of the five-character DTC code is a letter that indicates which of the car's systems the error code relates to. It can be about one of the following:

- Bodywork
- Chassis
- Powertrain
- Network

The next character in the sequence is either a 0 or a 1. This number indicates whether the

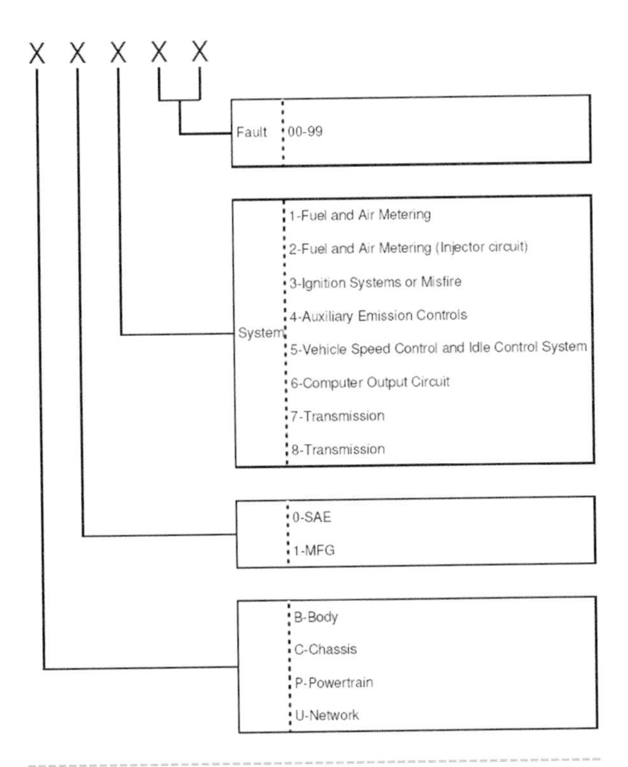

Figure 13-6 *DTC contents (Courtesy AutoTap)*

code is a generic code that applies to all vehicles or a specific problem with that type of vehicle.

0 = SAE generic fault code

1 = Manufacturer-specific fault code

The next digit carries the information as to what area of the power train control module the code refers to:

1 = Fuel and air metering

2 = Fuel and air metering (fuel injection system faults only)

3 = Ignition system or misfire monitor

4 = Auxiliary emission control system (exhaust gas recirculation, catalyst, and EVAP systems)

5 = Vehicle speed control and engine idle control

6 = Computer output circuits

7 = Transmission

8 = Transmission

9 = Transmission

The final two digits give the exact nature of the fault. The following is a list of generic DTC trouble codes you may encounter:

P0001—Fuel Volume Regulator Control Circuit/Open

P0002—Fuel Volume Regulator Control Circuit Range/Performance

P0003—Fuel Volume Regulator Control Circuit Low

P0004—Fuel Volume Regulator Control Circuit High

P0005—Fuel Shutoff Valve "A" Control Circuit Open

P0006—Fuel Shutoff Valve "A" Control Circuit Low

P0007—Fuel Shutoff Valve "A" Control Circuit High

P0008—Engine Position System Performance Bank 1

P0009—Engine Position System Performance Bank 2

P0010—"A" Camshaft Position Actuator Circuit Bank 1

P0011—"A" Camshaft Position—Timing Over—Advanced or System Performance Bank 1

P0012—"A" Camshaft Position—Timing Over—Retarded Bank 1

P0013—"B" Camshaft Position—Actuator Circuit Bank 1

P0014—"B" Camshaft Position—Timing Over—Advanced or System Performance Bank 1

P0015—"B" Camshaft Position—Timing Over—Retarded Bank 1

P0016—Crankshaft Position—Camshaft Position Correlation Bank 1 Sensor A

P0017—Crankshaft Position—Camshaft Position Correlation Bank 1 Sensor B

P0018—Crankshaft Position—Camshaft Position Correlation Bank 2 Sensor A

P0019—Crankshaft Position—Camshaft Position Correlation Bank 2 Sensor B

P0020—"A" Camshaft Position Actuator Circuit Bank 2

P0021—"A" Camshaft Position—Timing Over—Advanced or System Performance Bank 2

P0022—"A" Camshaft Position—Timing Over—Retarded Bank 2

P0023—"B" Camshaft Position—Actuator Circuit Bank 2

P0024—"B" Camshaft Position—Timing Over—Advanced or System Performance Bank 2

P0025—"B" Camshaft Position—Timing Over—Retarded Bank 2

P0026—Intake Valve Control Solenoid Circuit Range/Performance Bank 1

P0027—Exhaust Valve Control solenoid Circuit Range/Performance Bank 1

P0028—Intake Valve Control Solenoid Circuit Range/Performance Bank 2

P0029—Exhaust Valve Control Solenoid Circuit Range/Performance Bank 2

P0030—HO_2S Heater Control Circuit Bank 1 Sensor 1

P0031—HO_2S Heater Circuit Low Voltage Bank 1 Sensor 1

P0032—HO_2S Heater Circuit High Voltage Bank 1 Sensor 1

P0033—Turbo Charger Bypass Valve Control Circuit

P0034—Turbo Charger Bypass Valve Control Circuit Low

P0035—Turbo Charger Bypass Valve Control Circuit High

P0036—HO_2S Heater Control Circuit Bank 1 Sensor 2

P0037—HO_2S Heater Circuit Low Voltage Bank 1 Sensor 2

P0037—HO_2S Heater Control Circuit Low Bank 1 Sensor 2

P0038—HO_2S Heater Control Circuit High Bank 1 Sensor 2

P0039—Turbo/Super Charger Bypass Valve Control Circuit Range/ Performance

P0040—O2 Sensor Signals Swapped Bank 1 Sensor 1 and Bank 2 Sensor 1

P0043—HO_2S Heater Control Circuit Low (Bank 1, Sensor 3)

P0044—HO_2S Heater Control Circuit High (Bank 1, Sensor 3)

P0050—HO_2S Heater Control Circuit Bank 2 Sensor 1

P0051—HO_2S Heater Control Circuit Low Bank 2 Sensor 1

P0052—HO_2S Heater Control Circuit High Bank 2 Sensor 1

P0053—HO_2S Heater Resistance Bank 1 Sensor 1

P0054—HO_2S Heater Resistance Bank 1 Sensor 2

P0055—HO_2S Heater Resistance Bank 1 Sensor 3

P0056—HO_2S Heater Control Circuit Bank 2 Sensor 2

P0057—HO_2S Heater Control Circuit Low Bank 2 Sensor 2

P0058—HO_2S Heater Control Circuit High Bank 2 Sensor 2

P0059—HO_2S Heater Resistance Bank 2 Sensor 1

P0060—HO_2S Heater Resistance Bank 2 Sensor 2

P0061—HO_2S Heater Resistance Bank 2 Sensor 3

P0062—HO_2S Heater Control Circuit Bank 2 Sensor 3

P0063—HO_2S Heater Control Circuit Low Bank 2 Sensor 3

P0064—HO_2S Heater Control Circuit Low Bank 2 Sensor 3

P0065—Air Assisted Injector Control Range/Performance

P0066—Air Assisted Injector Control Circuit Low

P0067—Air Assisted Injector Control Circuit High

P0068—Manifold Absolute Pressure / Mass Air Flow (MAP/MAF)—Throttle Position Correlation

P0069—Manifold Absolute Pressure—Barometric Pressure Correlation

P0070—Ambient Air Temperature Sensor Circuit

P0071—Ambient Air Temperature Sensor Range/Performance

P0072—Ambient Air Temperature Sensor Circuit Low

P0073—Ambient Air Temperature Sensor Circuit High

P0074—Ambient Air Temperature Sensor Intermittent

P0075—Intake Valve Control Solenoid Circuit Bank 1

P0076—Intake valve Control Solenoid Circuit Low Bank 1

P0077—Intake Valve Control Solenoid Circuit High Bank 1

P0078—Exhaust Valve Control Solenoid Circuit Bank 1

P0079—Exhaust Valve Control Solenoid Circuit Low Bank 1

P0080—Exhaust Valve Control Solenoid Circuit High Bank 1

P0081—Intake Valve Control Solenoid Circuit Bank 2

P0082—Intake Valve Control Solenoid Circuit Low Bank 2

P0083—Intake Valve Control Solenoid Circuit High Bank 2

P0084—Exhaust Valve Control Solenoid Circuit Bank 2

P0085—Exhaust Valve Control Solenoid Circuit Low Bank 2

P0086—Exhaust Valve Control Solenoid Circuit High Bank 2

P0087—Fuel Rail/System Pressure—Too Low

P0088—Fuel Rail/System Pressure—Too High

P0089—Fuel Pressure Regulator 1 Performance

P0090—Fuel Pressure Regulator 1 Control Circuit

P0091—Fuel Pressure Regulator 1 Control Circuit Low

P0092—Fuel Pressure Regulator 1 Control Circuit High

P0093—Fuel System Leak Detected—Large Leak

P0094—Fuel System Leak Detected—Small Leak

P0095—Intake Air Temperature Sensor 2 Circuit

P0096—Intake Air Temperature Sensor 2 Circuit Range/Performance

P0097—Intake Air Temperature Sensor 2 Circuit Low

P0098—Intake Air Temperature Sensor 2 Circuit High

P0099—Intake Air Temperature Sensor 2 Circuit Intermittent/Erratic

P0100—Mass Air Flow (MAF) Sensor Circuit Insufficient Activity

P0101—Mass Air Flow System Performance

P0102—Mass Air Flow Sensor Circuit Low Frequency

P0103—Mass Air Flow Sensor Circuit High Frequency

P0104—Mass Air Flow Circuit Intermittent

P0105—Manifold Absolute Pressure Sensor Circuit Insufficient Activity

P0106—Manifold Absolute Pressure System Performance

P0107—Manifold Absolute Pressure Sensor Circuit Low Voltage

P0108—Manifold Absolute Pressure Sensor Circuit High Voltage

P0109—Manifold Absolute Pressure Circuit Intermittent

P0110—Intake Air Temperature Circuit

P0111—Intake Air Temperature Circuit Range/Performance

P0112—Intake Air Temperature Sensor Circuit Low Voltage

P0113—Intake Air Temperature Sensor Circuit High Voltage

P0114—Intake Air Temperature Circuit Intermittent

P0115—Engine Coolant Temperature Sensor Circuit

P0116—Engine Coolant Temperature Sensor Circuit Performance

P0117—Engine Coolant Temperature Sensor Circuit Low Voltage

P0118—Engine Coolant Temperature Sensor Circuit High Voltage

P0119—Engine Coolant Temperature Circuit Intermittent

P0120—Throttle Position Sensor Circuit

P0121—Throttle Position Sensor A Circuit Range/Performance

P0122—Throttle Position Sensor Circuit Low Voltage

P0123—Throttle Position Sensor Circuit High Voltage

P0124—Throttle Position Sensor A Intermittent

P0125—Engine Coolant Temperature Excessive Time to Closed Loop Fuel Control

P0126—Insufficient Engine Coolant Temperature for Stable Operation

P0127—Intake Air Temperature Too High

P0128—Coolant Thermostat (Coolant Temp Below Thermostat Regulating Temperature)

P0130—HO_2S Circuit Bank 1 Sensor 1

P0131—HO_2S Circuit Low Voltage Bank 1 Sensor 1

P0132—HO_2S Circuit High Voltage Bank 1 Sensor 1

P0133—HO_2S Slow Response Bank 1 Sensor 1

P0134—HO_2S Circuit Insufficient Activity Bank 1 Sensor 1

P0135—HO_2S Heater Circuit Bank 1 Sensor 1

P0137—HO_2S Circuit Low Voltage Bank 1 Sensor 2

P0138—HO_2S Circuit High Voltage Bank 1 Sensor 2

P0139—HO_2S Slow Response Bank 1 Sensor 2

P0140—HO_2S Circuit Insufficient Activity Bank 1 Sensor 2

P0141—HO_2S Heater Circuit Bank 1 Sensor 2

P0142—HO_2S Circuit Bank 1 Sensor 3

P0143—HO_2S Circuit Low Voltage Bank 1 Sensor 3

P0144—HO_2S Circuit High Voltage Bank 1 Sensor 3

P0145—HO_2S Circuit Bank 1 Sensor 2 Slow Response

P0146—HO_2S Circuit Insufficient Activity Bank 1 Sensor 3

P0147—HO_2S Heater Circuit Bank 1 Sensor 3

P0150—HO_2S Circuit Bank 2 Sensor 1

P0151—HO_2S Circuit Low Voltage Bank 2 Sensor 1

P0152—HO_2S Circuit High Voltage Bank 2 Sensor 1

P0153—HO_2S Slow Response Bank 2 Sensor 1

P0154—HO_2S Circuit Insufficient Activity Bank 2 Sensor 1

P0155—Heated Oxygen Sensor Heater Circuit (Bank 2, Sensor 1)

P0156—HO_2S Circuit Bank 2 Sensor 2

P0157—HO_2S Circuit Low Voltage Bank 2 Sensor 2

P0158—HO_2S Circuit High Voltage Bank 2 Sensor 2

P0159—HO_2S Slow Response Bank 2 Sensor 2

P0160—HO_2S Circuit Insufficient Activity Bank 2 Sensor 2

P0161—HO_2S Heater Circuit Bank 2 Sensor 2

P0162—HO_2S Circuit Bank 2 Sensor 3

P0163—HO_2S Circuit Bank 2 Sensor 3 Low Voltage

P0164—HO_2S Circuit Bank 2 Sensor 3 High Voltage

P0165—HO_2S Circuit Bank 2 Sensor 3 Slow Response

P0166—HO_2S Circuit Bank 2 Sensor 3 No Activity

P0167—HO_2S Heater Circuit Bank 2 Sensor 3

P0169—Fuel Composition Sensor

P0170—Fuel Trim Bank 1

P0171—Fuel Trim System Lean Bank 1

P0172—Fuel Trim System Rich Bank 1

P0174—Fuel Trim System Lean Bank 2

P0175—Fuel Trim System Rich Bank 2

P0176—Fuel Composition Sensor Circuit

P0177—Fuel Composition Sensor Circuit Performance

P0178—Fuel Composition Sensor Circuit Low Voltage

P0179—Fuel Composition Sensor Circuit High Voltage

P0180—Fuel Temperature Sensor A Circuit

P0181—Fuel Temperature Sensor A Circuit Range/Performance

P0182—Fuel Temperature Sensor A Circuit Low Input

P0183—Fuel Temperature Sensor A Circuit High Input

P0184—Fuel Temperature Sensor 1 Circuit Intermittent

P0185—Fuel Temperature Sensor 2 Circuit

P0186—Fuel Temperature Sensor B Circuit Range/Performance

P0187—Fuel Temperature Sensor B Circuit Low Input

P0189—Fuel Temperature Sensor 2 Circuit Intermittent

P0190—Fuel Rail Pressure Sensor Circuit

P0191—Fuel Rail Pressure Sensor Circuit Performance

P0192—Fuel Rail Pressure Sensor Circuit Low Voltage

P0193—Fuel Rail Pressure Sensor Circuit High Voltage

P0194—Fuel Rail Pressure Sensor Circuit Intermittent

P0195—Engine Oil Temperature Sensor Circuit

P0196—Engine Oil Temperature Sensor Performance

P0197—Engine Oil Temperature Sensor Low Voltage

P0198—Engine Oil Temperature Sensor High Voltage

P0199—Engine Oil Temperature Sensor Intermittent

P0200—Injector Control Circuit Voltage

P0201—Injector 1 Control Circuit

P0202—Injector 2 Control Circuit

P0203—Injector 3 Control Circuit

P0204—Injector 4 Control Circuit

P0205—Injector 5 Control Circuit

P0206—Injector 6 Control Circuit

P0207—Injector 7 Control Circuit

P0208—Injector 8 Control Circuit

P0209—Injector 9 Control Circuit

P0210—Injector 10 Control Circuit

P0211—Injector 11 Control Circuit

P0212—Injector 12 Control Circuit

P0213—Cold Start Injector 1

P0214—Cold Start Injector 2

P0215—Engine Shutoff Control Circuit

P0216—Injection Timing Control Circuit

P0217—Engine Over Temperature—Hot Light Requested

P0218—Transmission Fluid Over Temperature

P0219—Engine Overspeed Condition

P0220—Throttle Position Sensor 2 Circuit

P0222—Throttle Position Sensor B Circuit Low Voltage

P0223—Throttle Position Sensor B Circuit High Input

P0224—Throttle Position Sensor B Circuit Intermittent

P0225—Throttle Position Sensor 3 Circuit

P0226—Throttle Position Sensor 3 Circuit Performance

P0227—Throttle/Pedal Position Sensor/Switch "C Circuit Low"

P0228—Throttle/Pedal Position Sensor/Switch "C Circuit High"

P0228—APP Sensor 3 Circuit High Voltage

P0229—Throttle Position Sensor C Circuit Intermittent

P0230—Fuel Pump Relay Control Circuit

P0231—Fuel Pump Feedback Circuit Low Voltage

P0232—Fuel Pump Feedback Circuit High Voltage

P0234—Turbocharger Engine Overboost

P0235—Turbocharger Boost Sensor 1 Circuit

P0236—Turbocharger Boost Sensor 1 Performance

P0237—Turbocharger Boost Sensor 1 Circuit Low Voltage

P0238—Turbocharger Boost Sensor 1 Circuit High Voltage

P0239—Turbocharger Boost Sensor 2 Circuit

P0240—Turbocharger Boost Sensor 2 Performance

P0241—Turbocharger Boost Sensor 2 Circuit Low Voltage

P0242—Turbocharger Boost Sensor 2 Circuit High Voltage

P0243—Turbocharger Boost Solenoid Control Circuit

P0244—Turbocharger Wastegate Solenoid 1 Performance

P0245—Turbocharger Wastegate Solenoid 1 Low Voltage

P0246—Turbocharger Wastegate Solenoid 1 High Voltage

P0247—Turbocharger Wastegate Solenoid 2

P0248—Turbocharger Wastegate Solenoid 2 Performance

P0249—Turbocharger Wastegate Solenoid 2 Low Voltage

P0250—Turbocharger Wastegate Solenoid 2 High Voltage

P0251—Injection Pump Fuel Metering Control A

P0252—Injector Pump 1 Rotor/Cam Performance

P0253—Injector Pump 1 Rotor/Cam Low Voltage

P0254—Injector Pump 1 Rotor/Cam High Voltage

P0255—Injector Pump 1 Rotor/Cam Intermittent

P0256—Injector Pump 2 Rotor/Cam

P0257—Injector Pump 2 Rotor/Cam Performance

P0258—Injector Pump 2 Rotor/Cam Low Voltage

P0259—Injector Pump 2 Rotor/Cam High Voltage

P0260—Injector Pump 2 Rotor/Cam Intermittent

P0261—Cylinder 1 Injector Circuit Low

P0262—Cylinder 1 Injector Circuit High

P0263—Cylinder 1 Contribution/Balance

P0264—Cylinder 2 Injector Circuit Low

P0265—Cylinder 2 Injector Circuit High

P0266—Cylinder 2 Contribution/Balance

P0267—Cylinder 3 Injector Circuit Low

P0268—Cylinder 3 Injector Circuit High

P0269—Cylinder 3 Contribution/Balance

P0270—Cylinder 4 Injector Circuit Low

P0271—Cylinder 4 Injector Circuit High

P0272—Cylinder 4 Contribution/Balance

P0273—Cylinder 5 Injector Circuit Low

P0274—Cylinder 5 Injector Circuit High

P0275—Cylinder 5 Contribution/Balance

P0276—Cylinder 6 Injector Circuit Low

P0277—Cylinder 6 Injector Circuit High

P0278—Cylinder 6 Contribution/Balance

P0279—Cylinder 7 Injector Circuit Low

P0280—Cylinder 7 Injector Circuit High

P0281—Cylinder 7 Contribution/Balance

P0282—Cylinder 8 Injector Circuit Low

P0283—Cylinder 8 Injector Circuit High

P0284—Cylinder 8 Contribution/Balance

P0285—Injector Circuit Cylinder 9 Low Voltage

P0286—Injector Circuit Cylinder 9 High Voltage

P0287—Cylinder 9 Balance System

P0288—Injector Circuit Cylinder 10 Low Voltage

P0289—Injector Circuit Cylinder 10 High Voltage

P0290—Cylinder 10 Balance System

P0291—Injector Circuit Cylinder 11 Low Voltage

P0292—Injector Circuit Cylinder 11 High Voltage

P0293—Cylinder 11 Balance System

P0294—Injector Circuit Cylinder 12 Low Voltage

P0295—Injector Circuit Cylinder 12 High Voltage

P0296—Cylinder 12 Balance System

P0298—Engine Oil Overtemperature Condition

P0300—Random/Multiple Cylinder Misfire Detected

P0301—Cylinder 1 Misfire Detected

P0302—Cylinder 2 Misfire Detected

P0303—Cylinder 3 Misfire Detected

P0304—Cylinder 4 Misfire Detected

P0305—Cylinder 5 Misfire Detected

P0306—Cylinder 6 Misfire Detected

P0307—Cylinder 7 Misfire Detected

P0308—Cylinder 8 Misfire Detected

P0309—Cylinder 9 Misfire Detected

P0310—Cylinder 10 Misfire Detected

P0311—Cylinder 11 Misfire Detected

P0312—Cylinder 12 Misfire Detected

P0320—Ignition/Distributor Engine Speed Input Circuit

P0321—Ignition/Distributor Engine Speed Ckt. Performance

P0322—Ignition/Distributor Engine Speed Circuit No Signal

P0323—Ignition/Distributor Engine Speed Circuit Intermittent

P0325—Knock Sensor 1 Circuit Bank 1

P0326—Knock Sensor 1 Circuit Range/ Performance (Bank 1)

P0327—Knock Sensor 1 Circuit Low Input (Bank 1)

P0328—Knock Sensor 1 Circuit High Input (Bank 1)

P0329—Knock Sensor 1 Circuit Bank 1 Intermittent

P0330—Knock Sensor 2 Circuit Bank 2

P0331—Knock Sensor 2 Circuit Range/ Performance (Bank 2)

P0332—Knock Sensor 2 Circuit Low Input

P0333—Knock Sensor 2 Circuit High Input

P0334—Knock Sensor 2 Circuit Bank 2 Intermittent

P0335—Crankshaft Position (CKP) Sensor A Circuit

P0336—Crankshaft Position Sensor Circuit A Range/Performance

P0337—CKP Sensor Circuit Low Frequency

P0338—CKP Sensor Circuit High Frequency

P0339—CKP Sensor Circuit Intermittent

P0340—Camshaft Position Sensor A Circuit (Bank 1 or Single Sensor)

P0341—Camshaft Position Sensor A Circuit Range/Performance (Bank 1 or Single Sensor)

P0342—Camshaft Position Sensor Circuit Low Voltage

P0343—Camshaft Position Sensor Circuit High Voltage

P0344—Camshaft Position Sensor Circuit Intermittent

P0350—Ignition Coil Primary/Secondary Circuit

P0351—Ignition Coil A Primary/ Secondary Circuit

P0352—Ignition Coil B Primary/ Secondary Circuit

P0353—Ignition Coil C Primary/ Secondary Circuit

P0354—Ignition Coil D Primary/ Secondary Circuit

P0355—Ignition Coil E Primary/ Secondary Circuit

P0356—Ignition Coil F Primary/ Secondary Circuit

P0357—Ignition Coil G Primary/ Secondary Circuit

P0358—Ignition Coil H Primary/ Secondary Circuit

P0360—Ignition Coil J Primary/ Secondary Circuit

P0361—Ignition Coil K Primary/ Secondary Circuit

P0362—Ignition Coil L Primary/ Secondary Circuit

P0370—Timing Reference High Resolution System Performance

P0371—Too Many High Resolution Signal 1 Pulses

P0372—Too Few High Resolution Signal 1 Pulses

P0373—Intermittent High Resolution Signal 1 Pulse

P0374—No High Resolution Signal 1 Pulses

P0375—Timing Reference Signal 2 High Resolution

P0376—Too Many High Resolution Signal 2 Pulses

P0377—Too Few High Resolution Signal 2 Pulses

P0378—Intermittent High Resolution Signal 2 Pulse

P0380—Glow Plug/Heater Circuit A

P0381—Glow Plug/Heater Indicator Circuit

P0385—Crankshaft Position Sensor Circuit B

P0386—CKP Sensor B Circuit Performance

P0387—Crankshaft Position Sensor 2 Circuit Low Voltage

P0388—Crankshaft Position Sensor 2 Circuit High Voltage

P0389—Crankshaft Position Sensor 2 Circuit Intermittent

P0400—Exhaust Gas Recirculation Flow

P0401—Exhaust Gas Recirculation Flow Insufficient Detected

P0402—Exhaust Gas Recirculation Flow Excessive Detected

P0403—Exhaust Gas Recirculation Control Circuit

P0404—Exhaust Gas Recirculation Control Circuit Range/Performance

P0405—Exhaust Gas Recirculation Sensor A Circuit Low

P0406—Exhaust Gas Recirculation Sensor A Circuit High

P0407—EGR Sensor 2 Circuit Low Voltage

P0408—EGR Sensor 2 Circuit High Voltage

P0410—Secondary Air Injection System

P0411—Secondary Air Injection Incorrect Upstream Flow Detected

P0412—Secondary Air Injection Switching Valve A Circuit

P0413—Secondary Air Injection Switching Valve A Circuit Open

P0414—Secondary Air Injection Switching Valve A Circuit Shorted

P0416—Secondary Air Injection Switching Valve B Circuit Open

P0417—Secondary Air Injection Switching Valve B Circuit Shorted

P0418—Secondary Air Injection System Relay A Circuit

P0419—Secondary Air Injection System Relay B Circuit

P0420—Catalyst System Efficiency Below Threshold (Bank 1)

P0421—Warm Up Catalyst Efficiency Below Threshold (Bank 1)

P0422—Main TWC Efficiency Bank 1 Below Threshold

P0423—Heated TWC Efficiency Bank 1 Below Threshold

P0424—Heated TWC Temperature Bank 1 Below Threshold

P0426—Catalyst Temperature Sensor Range/Performance (Bank 1)

P0427—Catalyst Temperature Sensor Low Input (Bank 1)

P0428—Catalyst Temperature Sensor High Input (Bank 1)

P0430—Catalyst System Low Efficiency Bank 2

P0431—Warm Up Catalyst Efficiency Below Threshold (Bank 2)

P0432—Main TWC Efficiency Bank 2 Below Threshold

P0433—Heated TWC Efficiency Bank 2 Below Threshold

P0434—Heated TWC Temperature Bank 2 Below Threshold

P0436—Catalyst Temperature Sensor Range/Performance (Bank 2)

P0437—Catalyst Temperature Sensor Low Input (Bank 2)

P0438—Catalyst Temperature Sensor High Input (Bank 2)

P0440—Evaporative Emission Control System

P0441—Evaporative Emission Control System Incorrect Purge Flow

P0442—Evaporative Emission Control System Leak Detected (Small Leak)

P0443—Evaporative Emission Control System Purge Control Valve Circuit

P0444—Evaporative Emission Control System Purge Control Valve Circuit Open

P0445—Evaporative Emission Control System Purge Control Valve Circuit Shorted

P0446—Evaporative Emission Control System Vent Control Circuit

P0447—EVAP Vent Valve Control Circuit Open

P0448—EVAP Vent Valve Control Circuit Shorted

P0449—EVAP Canister Vent Solenoid Valve Control Circuit

P0450—Evaporative Emission Control System Pressure Sensor

P0451—Evaporative Emission Control System Pressure Sensor Range/ Performance

P0452—Evaporative Emission Control System Pressure Sensor Low Input

P0453—Evaporative Emission Control System Pressure Sensor High Input

P0455—Evaporative Emission Control System Leak Detected (Gross Leak/No Flow)

P0456—Evaporative Emission Control System Leak Detected (Very Small Leak)

P0460—Fuel Level Sensor Circuit

P0461—Fuel Level Sensor Circuit Range/Performance

P0462—Fuel Level Sensor Circuit Low Voltage

P0463—Fuel Level Sensor Circuit High Voltage

P0464—Fuel Level Sensor Circuit Intermittent

P0465—Purge Flow Sensor Circuit

P0466—Purge Flow Sensor Circuit Performance

P0467—Purge Flow Sensor Circuit Low Voltage

P0468—Purge Flow Sensor Circuit High Voltage

P0469—Purge Flow Sensor Circuit Intermittent

P0470—Exhaust Pressure Sensor

P0471—Exhaust Pressure Sensor Range/ Performance

P0472—Exhaust Pressure Sensor Low Input

P0473—Exhaust Pressure Sensor High Input

P0474—Exhaust Pressure Sensor Circuit Intermittent

P0475—Exhaust Pressure Control Valve

P0476—Exhaust Pressure Control Valve Range/Performance

P0477—Exhaust Pressure Control Valve Circuit Low Voltage

P0478—Exhaust Pressure Control Valve High Input

P0479—Exhaust Pressure Control Valve Intermittent

P0480—Coolant Fan 1 Control Circuit

P0481—Coolant Fan Relay 2 Control Circuit

P0500—Vehicle Speed Sensor

P0501—Vehicle Speed Sensor Range/Performance

P0502—Vehicle Speed Sensor Circuit Low Input

P0503—Vehicle Speed Sensor Intermittent

P0505—Idle Control System

P0506—Idle Control System RPM Lower Than Expected

P0507—Idle Control System RPM Higher Than Expected

P0508—Idle Control System Circuit Low

P0509—Idle Control System Circuit High

P0510—Closed Throttle Position Switch

P0512—Starter Request Circuit Performance

P0522—Engine Oil Pressure Sensor Circuit Low Voltage

P0523—Engine Oil Pressure Sensor Circuit High Voltage

P0530—A/C Refrigerant Pressure Sensor Circuit

P0532—Air Conditioning Refrigerant Pressure Sensor Circuit Low Voltage

P0533—Air Conditioning Refrigerant Pressure Sensor Circuit High Voltage

P0534—Air Conditioner Refrigerant Charge Loss

P0541—Intake Air Heater Circuit Low

P0542—Intake Air Heater Circuit High

P0550—Power Steering Pressure Sensor Circuit

P0551—Power Steering Pressure Sensor Circuit Range/Performance

P0552—Power Steering Pressure Sensor Circuit Low Voltage

P0553—Power Steering Pressure Sensor Circuit High Voltage

P0554—Power Steering Pressure Sensor Circuit Intermittent

P0560—System Voltage

P0561—System Voltage Unstable

P0562—System Voltage Low

P0563—System Voltage High

P0565—Cruise Control ON Signal

P0566—Cruise Control OFF Signal

P0567—Cruise Control RESUME Signal

P0568—Cruise Control SET Signal

P0569—Cruise Control COAST Signal

P0570—Cruise Control ACCEL Signal

P0571—Cruise Control Brake Switch A Circuit

P0572—Cruise Brake Switch 1 Circuit Low Voltage

P0573—Cruise Brake Switch 1 Circuit High Voltage

P0574—Cruise Control System—Vehicle Speed Too High

P0600—Serial Communication Link

P0601—Internal Control Module Memory Check Sum Error

P0602—Powertrain Control Module Programming Error

P0603—Powertrain Control Module Keep Alive Memory (KAM) Error

P0604—Internal Control Module Random Access Memory Error

P0605—Powertrain Control Module Read Only Memory (ROM) Error

P0606—ECM/PCM Processor

P0608—Vehicle Speed Output Circuit

P0615—Starter Relay Control Circuit

P0620—Generator Control Circuit

P0621—GEN Lamp L Control Circuit

P0622—GEN Field F Control Circuit

P0636—Power Steering Control Circuit Low

P0637—Power Steering Control Circuit High

P0640—Intake Air Heater Control Circuit

P0645—A/C Clutch Relay Control Circuit

P0646—A/C Clutch Relay Circuit Low Voltage

P0647—A/C Clutch Relay Circuit High Voltage

P0650—Malfunction Indicator Lamp (MIL) Control Circuit

P0654—Engine Speed Output Circuit

P0656—Fuel Level Output Circuit

P0660—Intake Manifold Tuning Valve Control Circuit—Bank 1

P0661—Intake Manifold Tuning Valve Control Circuit Low—Bank 1

P0662—Intake Manifold Tuning Valve Control Circuit High—Bank 1

P0666—PCM/ECM/TCM Internal Temperature Sensor Circuit

P0670—Glow Plug Module Control Circuit

P0671—Cylinder 1 Glow Plug Circuit

P0672—Cylinder 2 Glow Plug Circuit

P0673—Cylinder 3 Glow Plug Circuit

P0674—Cylinder 4 Glow Plug Circuit

P0675—Cylinder 5 Glow Plug Circuit

P0676—Cylinder 6 Glow Plug Circuit

P0677—Cylinder 7 Glow Plug Circuit

P0678—Cylinder 8 Glow Plug Circuit

P0683—Glow Plug Control Module to PCM Communication Circuit

P0684—Glow Plug Control Module to PCM Communication Circuit Range/Performance

P0700—Transmission Control System (MIL Request)

P0701—Transmission Control System Performance

P0702—Transmission Control System Electrical

P0703—Brake Switch B Input Circuit

P0703—Brake Switch Circuit

P0704—Clutch Switch Input Circuit

P0705—Transmission Range Sensor Circuit (PRNDL Input)

P0706—Transmission Range Sensor Circuit Range/Performance

P0707—Transmission Range Sensor Circuit Low Input

P0708—Transmission Range Sensor Circuit High Input

P0709—Transmission Range Sensor Circuit Intermittent

P0710—Transmission Fluid Temperature Sensor Circuit

P0711—Transmission Fluid Temperature Sensor Circuit Range/Performance

P0712—Transmission Fluid Temperature Sensor Circuit Low Input

P0713—Transmission Fluid Temperature Sensor Circuit High Input

P0714—Transmission Fluid Temperature Sensor Circuit Intermittent

P0715—Turbine Shaft Speed Sensor Circuit

P0716—Input/Turbine Speed Sensor Circuit Performance

P0717—Input/Turbine Speed Sensor Circuit No Signal

P0718—Turbine Shaft Speed Sensor Circuit Intermittent

P0719—Brake Switch 2 Circuit Low Voltage

P0720—Output Shaft Speed Sensor Circuit

P0721—Output Shaft Speed Sensor Circuit Range/Performance

P0722—Output Shaft Speed Sensor Circuit No Signal

P0723—Output Shaft Speed Sensor Circuit Intermittent

P0724—Brake Switch 2 Circuit High Voltage

P0725—Engine Speed Input Circuit

P0726—Engine Speed Input Circuit Performance

P0727—Engine Speed Circuit No Signal

P0728—Engine Speed Input Circuit Intermittent

P0730—Incorrect Gear Ratio

P0731—Gear 1 Incorrect Ratio

P0732—Gear 2 Incorrect Ratio

P0733—Gear 3 Incorrect Ratio

P0734—Gear 4 Incorrect Ratio

P0735—Gear 5 Incorrect Ratio

P0736—Reverse Incorrect Ratio

P0740—Torque Converter Clutch Circuit

P0741—Torque Converter Clutch Circuit Performance Stuck Off

P0742—Torque Converter Clutch Circuit Stuck On

P0743—Torque Converter Clutch Circuit Electrical

P0744—Torque Converter Clutch Circuit Intermittent

P0745—Pressure Control Solenoid A

P0746—Pressure Control Solenoid A Performance or Stuck Off

P0747—Pressure Control Solenoid Stuck On

P0748—Pressure Control Solenoid Electrical

P0749—Pressure Control Solenoid Circuit Intermittent

P0750—Shift Solenoid A

P0751—Shift Solenoid A Performance or Stuck Off

P0752—Shift Solenoid A Stuck On

P0753—Shift Solenoid A Electrical

P0754—1–2 Shift Solenoid Intermittent

P0755—Shift Solenoid B

P0756—Shift Solenoid B Performance or Stuck Off

P0757—Shift Solenoid B Stuck On

P0758—Shift Solenoid B Electrical

P0761—Shift Solenoid C Performance or Stuck Off

P0762—3–4 Shift Solenoid Stuck On

P0763—Shift Solenoid C Electrical

P0764 —3–4 Shift Solenoid Intermittent

P0765—Shift Solenoid D

P0766—Shift Solenoid D Performance or Stuck Off

P0767—4–5 Shift Solenoid Stuck On

P0768—Shift Solenoid D Electrical

P0769—4–5 Shift Solenoid Intermittent

P0770—Shift Solenoid E

P0772—Shift Solenoid E Stuck On

P0773—Shift Solenoid E Electrical

P0774—Shift Solenoid 5 Intermittent

P0775—Pressure Control Solenoid B

P0779—Pressure Control Solenoid B Intermittent

P0780—Shift

P0781—1–2 Shift

P0782—2–3 Shift

P0783—3–4 Shift

P0784—4–5 Shift

P0785—Shift Timing Solenoid

P0786—Shift Timing Solenoid Performance

P0787—Shift Timing Solenoid Low Voltage

P0788—Shift Timing Solenoid High Voltage

P0789—Shift Timing Solenoid Intermittent

P0790—Normal/Performance Switch Circuit

P0791—Intermediate Shaft Speed Sensor Circuit

P0794—Intermediate Shaft Speed Sensor Circuit Intermittent

P0795—Pressure Control Solenoid C

P0796—Pressure Control Solenoid C Performance or Stuck Off

P0797—Pressure Control Solenoid C Stuck On

P0799—Pressure Control Solenoid C Intermittent

P0801—Reverse Inhibit Control Circuit

P0803—1–4 Upshift (Skip Shift) Solenoid Control Circuit

P0804—1–4 Upshift (Skip Shift) Lamp Control Circuit

P0812—Reverse Input Circuit

P0814—Transmission Range Display Circuit

P0815—Upshift Switch Circuit

P0816—Downshift Switch Circuit

P0818—Driveline Disconnect Switch Input Circuit

P0840—Transmission Fluid Pressure Sensor/Switch A Circuit

P0841—Transmission Fluid Pressure Sensor/Switch A Circuit Range/ Performance

P0844—Transmission Fluid Pressure Sensor/Switch A Circuit Intermittent

P0846—Transmission Fluid Pressure Sensor/Switch B Circuit Range/ Performance

Chapter Fourteen

Wireless Connections

Imagine the possibilities and doors that are opened by wirelessly networking your car PC. Internet on the move becomes a possibility in public wireless hotspots. Media can be seamlessly transferred from the home PC to the car without the need for cables. The possibility of making telephone calls using VoIP becomes a reality. Data can be seamlessly transferred between laptop, PDA, and car PC.

This chapter is not intended to be an exhaustive guide to wireless networking, but merely an introduction to the possibilities of connecting your car PC to other PCs and networks. We will examine the different technologies available for connecting devices to your car PC wirelessly and evaluate their suitability for different tasks.

Wireless connections can be obtained using a variety of technologies:

- IrDa (Infrared)
- Bluetooth
- GSM
- GPRS
- 3G
- WiFi, WLAN, 802.11x

Let's address these technologies separately.

Infrared

Infrared is suitable for small-distance, slow-speed communications where the sending and receiving devices are in line of sight. This technology is used in the cheaper wireless keyboard systems and also to connect PDAs, laptops, and mobile phones to the car PC. I recommend infrared communications only for devices inside the car.

Most infrared keyboards and remote controls come with a proprietary infrared interface and will not work with an IrDa interface within the normal Windows set-up for IrDa communications.

Check this Web site for a range of infrared decoder hardware and software: www.evation.com/irman/

Using the IrMan software, you can create macros to execute commonly performed sets of instructions at the push of a remote control button. Furthermore, you can use supplied plug-ins to allow you to control many common PC programs such as WinAmp and Real Jukebox. The software also comes with Linux support.

Project 48: Building an IrDa Interface

Many motherboards have a connector for an infrared interface module. These modules can be difficult to find preassembled and are often very expensive for what is only a handful of components.

In this project we will assemble an IrDa interface module. This is the hardware side of things. You will need to find appropriate software for both the transmitting and receiving devices.

Look for a connector labeled with one of the following on your motherboard: IR, IRCON, SIR, or SIRCON. As many people will be using the VIA Mini-ITX range of motherboards, I have included a diagram of the location and orientation of the connector in Figure 14-1. The pinout is given in Table 14-1.

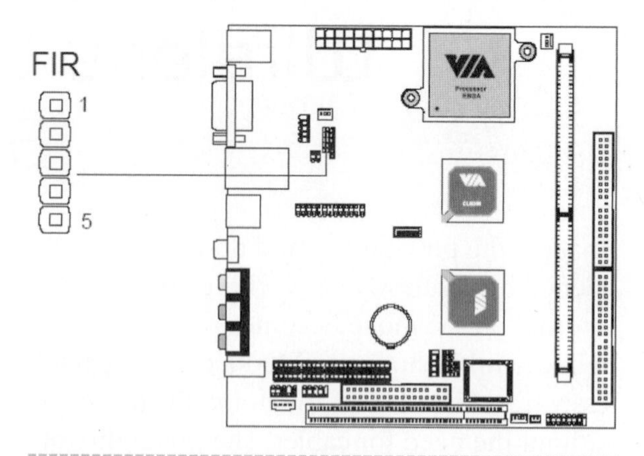

Figure 14-1 *Location and orientation of the connector (Courtesy VIA)*

Table 14-1

Pinouts

Pin	*Signal*
1	Vcc
2	Ir Receive (Rx)
3	Ir Receive (Rx) 2
4	GND
5	Ir Transmit (TX)

You Will Need

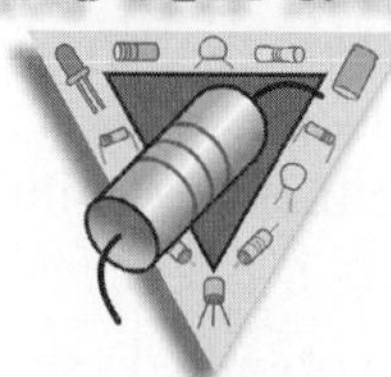

- Infrared LED
- Infrared photodiode
- BC 548 transistors (×2)
- 10 μF capacitor
- 47K resistor
- 47K resistor
- 15K resistor
- 22R resistor
- 1K resistor
- Small piece of stripboard
- Length of header socket
- Ribbon cable
- Heatshrink

Tools

Soldering iron

Craft knife

Hot-melt glue gun (optional)

Heat gun (optional)

Assemble the circuit to the diagram in Figure 14-2. Make sure that both the LED and photodiode are aligned to point in the same direction. A little hot-melt glue will secure them in place.

Solder the short length of ribbon cable to the various connection points.

Now take the length of female header and, using a craft knife, cut it into individual sockets. Solder each of these sockets to the opposite ends of the ribbon cable. Using the heat gun and shrinkwrap, make sure each of these individual wires is insulated so they do not short.

And that is about it! All you do now is connect the device to your motherboard in accordance with the pinout provided in your instruction manual. The four wires will be labeled Vcc, GND, TX, and RX.

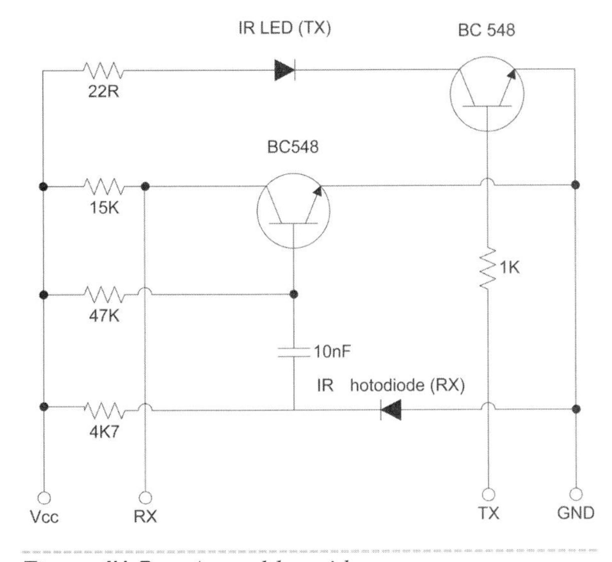

Figure 14-2 *Assembly guide*

You will need to set the device to full duplex mode in the motherboard setup to ensure you are using the device at best efficiency.

Windows will automatically detect the device and display the infrared logo in the taskbar.

Bluetooth

Bluetooth is not a technology that is suited to transmitting large amounts of data over a long distance. Bluetooth is better suited to transferring small pieces of information, such as address book details, pictures, and short media clips between a car PC and mobile phone or PDA.

Bluetooth is a good technology for connecting your car PC to peripheral devices without the need for messy cables. GPS mice are now available that interface with Bluetooth. This has the advantage that you can use the mouse with your car PC, laptop, and PDA without having to keep an array of cables.

To use Bluetooth devices with your PC, you need to have a Bluetooth dongle installed. These are readily available quite inexpensively, commonly connecting to a spare USB port. Bluetooth devices are small and inconspicuous; a model from D-Link is illustrated in Figure 14-3.

Figure 14-3 *D-Link Bluetooth Device*

GSM

GSM stands for *global system for mobile communications.* It is the technology that drove the first digital mobile phone networks and is widely used with over 1 billion subscribers.

With a PC connected to a GSM modem, your car PC will be able to connect to the internet in over 170 countries. GSM is a second-generation mobile technology, the successor to analogue mobile phones. *General packet radio service* (GPRS) is to be preferred to GSM where available.

Using Your Mobile Phone in Your Car

You can also use your car PC to control your mobile phone allowing you to use your PC as an extension of the phone to hold commonly used numbers and make text message input easier.

A piece of software called Phone Control allows you to manage all of the functions of your mobile phone from your car PC. Its features include the following:

- Fully skinnable user configurable interface (you can change how it looks onscreen by applying a custom "skin" with your own graphics or customize button placement and so on)
- Ability to connect to your phone using an IrDa Bluetooth or cable connection
- Touchscreen support
- Ability to dial numbers direct from phone book or manually
- Ability to send and receive text messages
- Ability to display mobile battery status
- Ability to mute media player while using phone

This software is ideally placed for use with a car PC as it fully integrates with other software and gives a professional appearance.

Go to www.phonecontrol.net to download this great software!

GPRS

General packet radio service, sometimes referred to as "2.5G" is a leap ahead for wireless Internet access. Modules that support GPRS communication will also support GSM as a rule, using the most efficient standard where it is available. GPRS can be used with mobile phones that support the feature to allow faster Internet access than is possible with GSM.

3GSM (3G)

Third-generation global system for mobile communications, or 3G, is the system used in all new 3G mobile phones. 3G opens up the possibility of wireless broadband with connection speeds many, many times faster than the original GSM system. 3G access is heavily dependent on regional availability. Connection to a car PC is done using either a 3G mobile phone with a lead or a data card similar to those used for GSM/GPRS.

WiFi/Wireless LAN/802.11x

Wireless fidelity (WiFi), also known as Wireless LAN or denoted by the standard specification 802.11x, is a method for networking computers without wires. It enables you to perform functions such as sharing computers, printers, or an internet connection, without the need for a physical connection.

In car PCs, a wireless connection can be invaluable as it would allow you to transfer media from your home PC to the car seamlessly without having to trail long network cable out to the garage every time you want a change of music on your hard drive.

The advantage of a WiFi connection over other wireless communication methods is that it allows for incredibly fast data transfer rates, much higher than can be achieved with other technologies. The speed of a WiFi modem is many times faster than via broadband or cable, making streaming video and media content a reality.

Many public locations such as bars, cafes, hotels, conference centers, and even gas stations are becoming WiFi enabled, meaning any computer (including a car PC) within range of that network can access the services provided. Some of these services are provided free as a complimentary bonus for the customers using the facility. Other WiFi services are operated by pay-as-you-go operators who require a payment up front for access.

Here are some good places to start looking for wireless hotspots:
www.wifinder.com/
www.wi-fihotspotlist.com/
www.wi-fizone.org/zoneFinder.asp?TID=7
www.wififreespot.com/europe.html

The family of specifications that are used for wireless networking is described in the following paragraphs.

The 802.11 Wireless Networking Specification Family

802.11 is a family of wireless networking technologies. Specific different technologies are denoted by a suffixed letter.

- 802.11 (the original standard)—The 2.4 GHz band provides 1 or 2 Mbps transmission rates.
- 802.11a—provides up to 54 Mbps in the 5 GHz band. 802.11a uses an orthogonal frequency division multiplexing encoding scheme and, as such, differs from the other two in the family.
- 802.11b—operates in the 2.4 GHz band and provides 11 Mbps transmission with the provision to fallback to 5.5, 2, and 1 Mbps when conditions are less favorable.
- 802.11g—also operates in the 2.4 GHz band and provides 20+ Mbps transmission rates.

It is important to note that the quoted speeds are the theoretical maximum speeds; results will vary depending on weather conditions, landscape topology, and a wide variety of variables. These external factors will also affect the distance that signals can be transmitted.

You may find, if your car is parked a long way away from your house or if something physical is obstructing the line of sight of the signal, you need to extend the range of your wireless networking setup. If you want to increase the range of your wireless networking capability, it is possible to add an external antenna to most WiFi cards to enable you to fit an aftermarket (or homebuilt) additional antenna. The antenna needs to be tuned to receive the range of frequencies used in WiFi communication. Ordinary car antennas will not cut the mustard.

A simple antenna that has quickly gained popularity is the waveguide antenna, sold under the brand name Cantenna and illustrated in Figure 14-4.

The Cantenna is available from Wireless Garden at www.cantenna.com.

Antennas should be mounted outside the vehicle as the fabric of the vehicle's body will prevent effective reception of signals and at best will degrade performance of the connection.

How Does the Waveguide Antenna Work?

The high-frequency signal enters the Cantenna, bouncing off its walls until it hits the sealed end of the can where it is reflected back. As the two waves interfere with each other, the reflected wave can be added to the incoming wave in accordance with the principles of superposition to produce a resultant. A pattern of standing waves is created whereby the signal is amplified at some points in the can and attenuated at others due to the way the waves interfere.

A standing wave is produced whenever two waves of the same frequency are traveling through a medium in different directions and interfere with each other. There are points on this wave that undergo no displacement; these are referred to as nodes. In contrast, the points that undergo maximum displacement are referred to as antinodes. The N-type connector is positioned at one such antinode so that it undergoes maximum displacement, providing a strong signal.

Figure 14-4 *Cantenna waveguide antenna (Courtesy Wireless Garden)*

Project 49: Building a Cantenna

You Will Need

You Will Need
- N-type female connector (single-hole mounting style)
- Can, 7 to 10 cm in diameter and as long as possible
- Short length of thick wire
- Pigtail lead for connection to WiFi hardware

The aim of this project is to construct a waveguide antenna for the least amount of money possible. Constructing the Cantenna is incredibly simple but must be done accurately for the antenna to be effective. You will need to measure the diameter of your can and compare it to Table 14-2. This will give you important information about the properties of an antenna built with that size can. It will also help you to work out where to position the connector as this is a crucial part of the operation.

The N-type connector, available from RS Components (listed at the back of the book) or many other electronics vendors, requires a little modification before you can use it. You will need to solder a length of copper wire to the end of the connector. Figure 14-5 gives the correct dimensional details and shows how the connector mounts to the can.

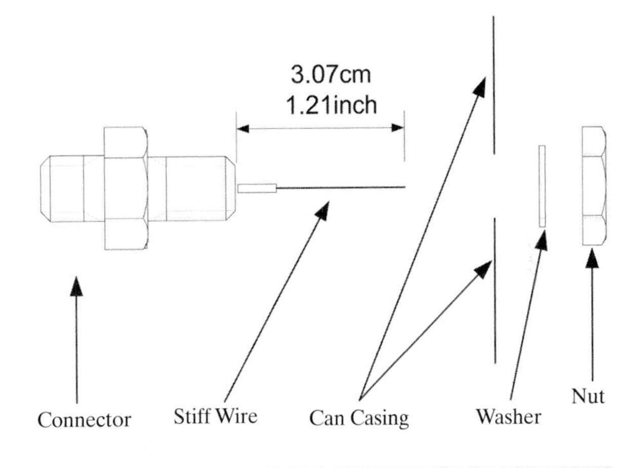

Figure 14-5 *Assembly Drawing for N-Type Connector*

Table 14-2

Critical dimensions for Cantenna construction

Inches	Centimeters	Cutoff Frequency (MHz) TI1 Mode	Cutoff Frequency (MHz) TM01 Mode	Guide Wavelength (inches)	Guide Wavelength (centimeters)	¼ Guide Wavelength (inches)	¼ Guide Wavelength (centimeters)	¾ Guide Wavelength (inches)	¾ Guide Wavelength (centimeters)
2.90	7.37	2,385.26	3,115.47	23.63	60.02	5.91	15.01	17.74	45.06
3.00	7.62	2,305.75	3,011.62	14.96	38.00	3.74	9.50	11.22	28.50
3.10	7.87	2,231.37	2,914.17	12.05	30.61	3.01	7.65	9.04	22.96
3.20	8.13	2,161.64	2,823.39	10.49	26.64	2.62	6.65	7.87	19.99
3.30	8.38	2,096.14	2,737.83	9.50	24.13	2.38	6.05	7.13	18.11
3.40	8.64	2,034.49	2,657.31	8.80	22.35	2.20	5.59	6.60	16.76
3.50	8.89	1,976.36	2,581.39	8.28	21.03	2.07	5.26	6.21	15.77
3.60	9.14	1,921.46	2,509.68	7.87	19.99	1.97	5.00	5.90	14.99
3.70	9.40	1,869.53	2,441.85	7.55	19.18	1.89	4.80	5.66	14.38
3.80	9.65	1,820.33	2,377.59	7.28	18.49	1.82	4.62	5.46	13.87

Figure 14-6 shows how the hole needs to be cut to mount the N-type connector.

For an antenna to work effectively with equipment designed to meet the 802.11b and 802.11g standards, the figures for your can diameter should give a TE11 cut-off frequency of less than 2.412, and the TM01 cut-off figure should be greater than 2.462.

Using the measurements derived from the table, you need to position the hole for the connector at a point that is the same distance away from the closed end of the can as the $^1/_4$-guide wavelength. This will determine the *center* of your hole, not the edge. You need to cut the hole in accordance with Figure 14.6. It is best if your can is longer than the $^3/_4$ -guide wavelength for effective operation.

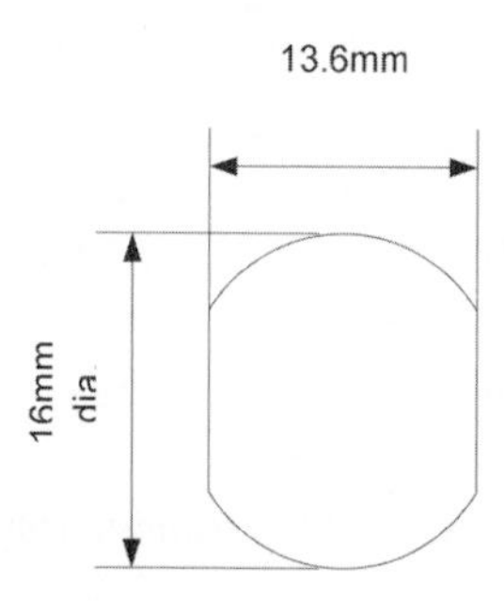

Figure 14-6 *Cutting the hole for the N-type connector*

You need to fix a cap to your Cantenna. This should be made of a plastic material that passes 2.4 GHz waves with minimal loss. Plastics that are particularly suitable are those used in microwaveable plastic, such as Tupperware.

To see if the material is suitable, a test can be performed with a microwave oven. Place the material in a domestic microwave oven alongside a cup of water. Let the microwave run for a few minutes until the water begins to boil. If you cannot detect a change in temperature of the plastic, then it should be suitable for use as a cap.

To connect your Cantenna to your car PC, you will need to obtain a lead commonly referred to as a "pigtail." This has an N-type connector on one end and a connector for your Cantenna on the other.

Commercially built Cantennas are available from www.cantenna.com.

Going Further . . .

If you are interested in a variety of aerial designs, the following enthusiasts' Web sites may be of interest to you:
www.wlan.org.uk/antenna-page.html
http://wb8erj.home.att.net/
www.wb8erj.com/
http://flakey.info/antenna/waveguide/
www.wlan.org.uk/tincan.gif
www.amsterdamwireless.net/workshop1512/3.html

Project 50: Going Wardriving

Wardriving is the process of driving around in a vehicle while using a car PC to find wireless access points that allow you to connect to the Internet.

You Will Need

You Will Need

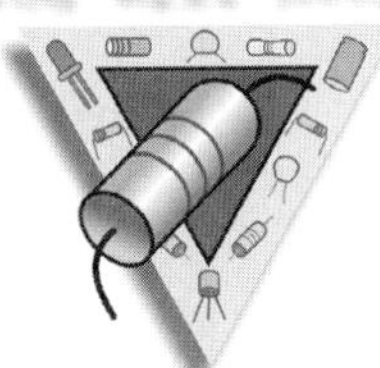

- A car
- Valid insurance and driver's license
- Car PC
- Wireless networking card
- Software
- GPS receiver (optional)

Warning

Legal Disclaimer: Before attempting to access any wireless network, you must ensure that you have permission to access that network or that it is in the public domain. Attempting to break encryption keys, use bandwidth without permission, or store files on a computer to which you do not have any rights is unethical and may be illegal in your locality.

With your car PC fired up and suitable software installed, wardriving is simply a matter of driving around your locality, looking for a wireless signal. Good places to look for are large offices, where stray wireless networks often "leak out" onto the highway. You might also like to try nearby schools, universities, and community centers. Many of these places offer free wireless access, not to mention libraries.

www.

www.netstumbler.com and www.netstumbler.org are great Web sites to keep you up to date with the movers and the shakers in the wireless networking world.

www.wardriving.com and www.wardriving.net both offer great portals to the world of wardriving. They give information on local hotspots as well as providing links to LOADS of useful software, much of it free and developed by the community.

Chapter Fifteen

Car PC Software

If you have followed this book up to this point, then the chances are you have now assembled a cracking car PC system, which is unparalleled in specification. The only thing is —it doesn't work. This is for the simple reason that you haven't installed any software yet! What follows is a guide to the software applications that can be used with your car PC to enhance your automotive environment.

Before you even start thinking about applications, you will need to give your system an operating system. Some VIA boards allow you to play media directly from boot up without having to use an operating system. This is okay for most basic needs; however, the chances are, after spending all that money on great kits, you want a little more functionality from your car PC!

In this chapter I am going to focus on installations using Windows XP. Although Microsoft products aren't to everyone's liking, they are understood by the vast majority of users and can be suited to a car PCs needs fairly easily.

That is not to say that car PC software is not out there for other platforms. A vast proliferation of Linux car PC software exists that runs well and is very stable. For Linux users a quick search on the Web will reveal a plethora of options available to the Linux user.

To make your car PC user friendly, you will need some front-end software. Front ends provide a way of tying all of your car PC applications together to give the user what appears to be a fully integrated car PC experience.

They tie up a bunch of disparate programs, such as GPS software, media players, and mobile phone controllers, and give them a pretty face. This has nice chunky buttons and features such as an onscreen keyboard (often referred to in forums as OSKs) to enable the user to have an enjoyable car PC experience.

CentraFuse

CentraFuse (see Figure 15-1) is one of the few pieces of car PC software that offers multilanguage support (available from www.fluxmedia.net). Some of its more salient features include the following:

- Audio mixer controls
- Customizable hot key support
- Id3-tag and directory mode
- Integrated playlists and favorites
- Integrated FM radio support (with radiator)

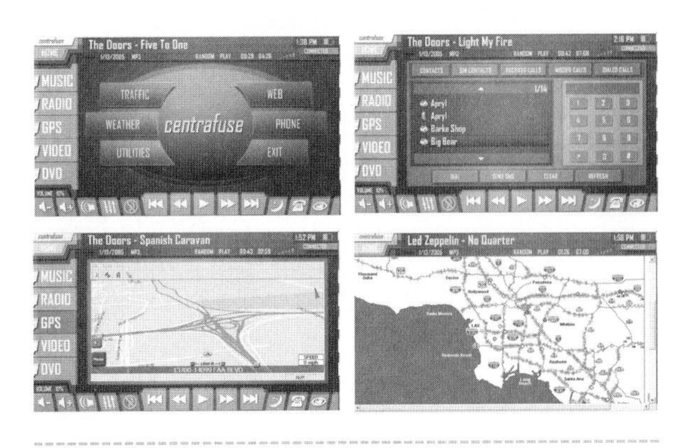

Figure 15-1 *CentraFuse screenshots (Courtesy www.mp3car.com)*

- Video playback (poster image support in video manager)
- Integrated DVD
- Internet status (LAN, GPRS, wireless)
- Ability to disconnect and connect default Internet connection
- Integrated weather and Web browser
- Free db with a local db for storage (used when CD inserted and ripping)
- Speech/voice recognition
- CD ripping
- Fully embedded GPS support
- Fully integrated phonecontrol .net

FrodoPlayer

FrodoPlayer is a nice piece of front-end software with support for both FM and XM radio (see Figure 15-2). DVD playback is supported as well as a rich array of media. Something that many users have commented they like about Frodoplayer is the video progress bar, which allows you to see how far into a media clip you are. Album art is displayed when you are playing a media file with an associated cover image.

Figure 15-2 *FrodoPlayer in car (Courtesy mp3car.com)*

Media Engine

Media Engine is one of the established heavyweights of the range of car PC front-end software (see Figure 15-3). It has become a staple software for most car PC diets in the early days. One of my only criticisms of Media Engine is that at times it can be a little sluggish. But other than that, I believe it is a very good quality product.

The even better news is . . . it is a free download.

Mobile Impact

Mobile Impact (see Figure 15-4) is a feature-rich front-end software that integrates the best features that a car PC can offer (http://mobileimpact.biz.tm).

The media player is based on Windows Media Player, which allows it to support a

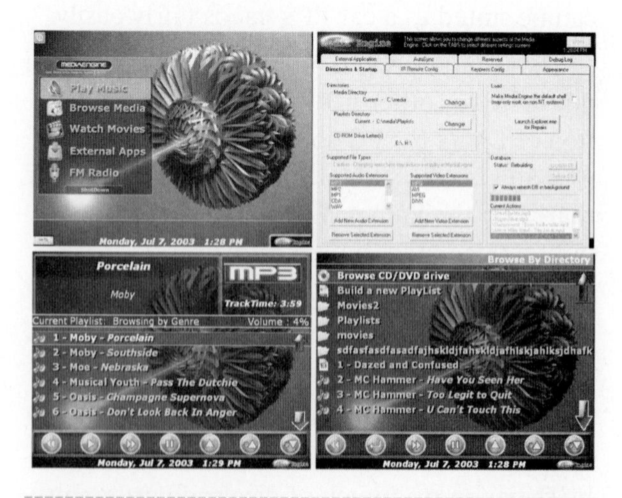

Figure 15-3 *Media Engine screenshots (Courtesy mp3car.com)*

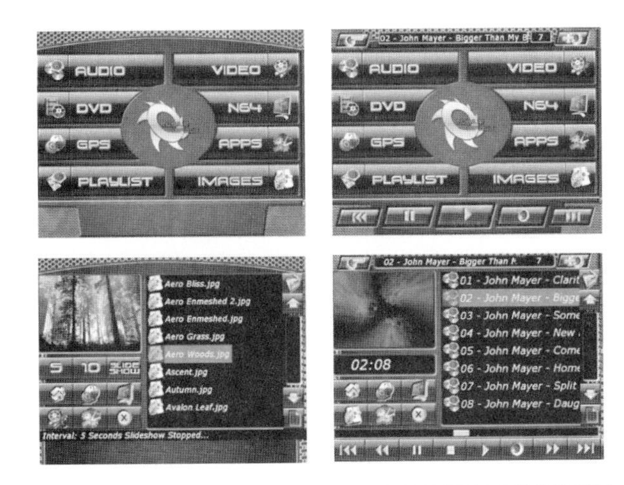

Figure 15-4 *Mobile Impact screenshots (Courtesy mp3car.com)*

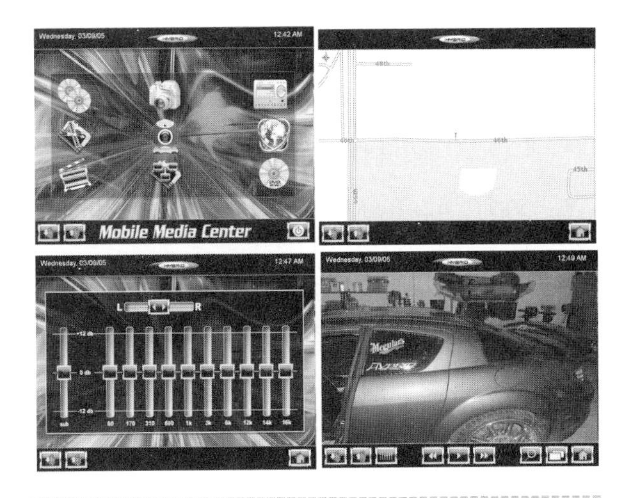

Figure 15-5 *Mobile Media Centre screenshots (Courtesy mp3car.com)*

wide range of file formats. The player will also play DVDs directly from disk, with no other external software required. As with all good front-end software, it has integrated functionality to support managing playlists. It also has a feature that allows you to create slideshows of your favorite images.

It has a number of other features. Unusually, the software also features a Nintendo 64 emulator to allow you to play games in the car. Any external GPS software is supported, as is any other external software programs. As appears to be standard fare with most front-end software, the interface is fully skinnable.

Mobile Media Centre

Mobile Media Centre (see Figure 15-5) is a well-specified front-end software that gives the user the option of voice control using a variety of spoken commands.

The software is easy to navigate using a touchscreen monitor and even comes with a scroll bar for navigating lists.

The software is incredibly fast at indexing media and presenting playlists to the user. This is a refreshing change from some of the more sluggish software.

A wide variety of media is supported in this software, with native support for MP1 through MP3, OGG, WMA, WAV, Redbook Audio (Audio CDs), MPG/MPEG, AVI/DIVX, QT/MOV, and WMV.

A nice feature is the video preview mode while browsing. When you eventually select a video file, you are offered the choice of three video play states (mini, medium, or full).

For DVD playback, the software uses its own codec that supports the following:

- NTSC mode 352×240, 352×480, 720×480, and PAL mode 352×288, 352×576, 720×576
- 24 fps or 30 fps DVD movie
- 4:3 and 16:9 aspect ratio
- 32, 44.1, and 48 KHz sample rate for AC-3 audio
- Mixing of 5.1 channel into two-channel Dolby surround stereo sound
- Sample rate 48 KHz and 96 KHz for linear PCM (LPCM)
- 16, 20, and 24 bits/sample
- MPEG-1 and MPEG-2 for MPEG audio
- Up to 32 subtitles, 52K biggest datapack and highlight button

- Up to eight languages
- Point-to-point loop function
- LINE21 output
- Video lightness and color adjustability
- Up to eight parent-control grades
- Multiple view angles, up to nine view angles
- High-performance deinterleave technology

The software also supports control of vehicle devices using an additional relay box, a reversing camera based on a webcam, and XM radio (for users in the United States).

GPS Software

We covered the installation of the GPS hardware in Chapter 12. We will now look at the array of options available for GPS software.

People often ask which is the best GPS software. The short answer is, there isn't one! Different GPS software will suit various users and their individual applications and setups. It is important to note that some software is designed to work only in the United States or Europe, and so it is important to check this before committing to buy anything.

Examine the features of your GPS software carefully. What follows is a brief guide to things that you should look out for.

Front-end embeddability is important in GPS software. Many packages come with an SDK, or software developer's kit. This allows talented individuals to write software programs that exploit the GPS engine, while providing a more-friendly interface for an automotive environment.

An important thing to remember is that most GPS software is designed to work on a desktop or laptop PC with a large screen and mouse as the input device. As a result, the icons are all very small when viewed on a smaller LCD touchscreen monitor. To overcome this problem, many front-end software interfaces come with a GPS plug-in that enables you to have big buttons and sometimes voice control to change the view of your GPS. This is friendlier for use in an in-car environment.

Some software packages are purely mapping software that will allow you to plot a route on a map using addresses or postal codes. For in-car use, software that incorporates turn-by-turn directions or guidance is to be preferred.

Something to look for in GPS software is a places-of-interest feature. These are coordinate locations on the map that can be turned on depending on user preferences. They allow you to locate utilities, services, and attractions quickly and easily. Better GPS software has a larger, more up-to-date database of places of interest.

When choosing GPS software, consider carefully what type of views you require. All GPS software will generally offer a two-dimensional bird's-eye view. This is like looking at a paper map. However, software packages exist that offer 3D views and various other permutations of the two.

GPS software packages derive their mapping information from a number of different providers. Some providers have a better reputation for the accuracy of their mapping information than others do. You may want to read some reviews on the Internet of different mapping providers before selecting the one whose mapping information best covers your area.

Some GPS software is better at finding a quick route than others. Generally, you are given two options when planning a route; you can specify the shortest or the quickest route. The shortest route will attempt to plot the route with the least number of miles between two points. The quickest route however, will take account of the amount of traffic on different roads at the time of day that you specify. This can result in a long protracted route that takes you out of the way on a scenic tour!

Rerouting is very useful if you take a wrong turn! GPS software should dynamically recalculate the next quickest route while you are on the move.

Note: Some route planning software does not take into consideration one-way streets particularly well. This may result in you having to go out of your way and allowing the software to reroute.

Destinator

Destinator is highly regarded in car PC circles and is available in various incarnations, both PC based and PDA based. The latest software provides great 3D views while still maintaining 2D functionality.

The software integrates with many front-end software packages. A car-PC-friendly interface is being developed known as "map-monkey," which is available with a variety of skins. The places-of-interest functionality that comes with Destinator is very easy to use. The map can be resized and rotated with ease, and the screen is updated and refreshed quickly. The software has the additional benefits brought by voice guidance. Finally a nice feature for touchscreen users is the onscreen keyboard.

One of the major gripes with Destinator is the fact that in 3D map mode, the street names are lost. This can be a little confusing. Further to this, the software works by dividing the maps into sections, which can make cross-country routing awkward.

For European users, Destinator comes with good European support.

iGuidance/Routis

iGuidance or Routis is another nice GPS application. It has the advantage over Destinator that its 3D views display street names. The software is touchscreen friendly. There is also an onscreen keyboard and voice guided directions.

Against iGuidance is the fact that some users have experienced problems when using a USB GPS device. The points-of-interest database can be confusing at the best of times and has a limited number of entries.

Delorme Street Atlas USA 2005

Delorme Street Atlas is available only to U.S. users. One of the main points in its favor is the exceptional quality of its maps. It also features turn-by-turn voice guidance and some voice control of its interface. Places of interest are dealt with in a unique manner using the POI Radar. The database is good with good entries in sufficient number. When driving and following a route, the map automatically rotates, which is a great help when trying to find your way. One of the features that makes this product stand out is the high contrast.

Unfortunately, use of this software is limited to users with bigger screens; the interface is not particularly touchscreen friendly due to the fact that the icons, tabs, buttons, and controls are all rather small. It does not have an onscreen keyboard, which can be restrictive when using a touchscreen. Integration with front-end software is also particularly difficult due to the way the software works.

Co-Pilot

Co-Pilot has good maps that are clear to read. The onscreen keyboard helps make this software touchscreen friendly. When driving, a helpful feature is the fact that the software zooms in on turns and junctions as you approach them. Particularly useful is the driver safety view, which places the distance of the next leg of the journey in large text when the vehicle is going over 10 mph.

What is nice about Co-Pilot is the fact that it is a guidance program rather than a mapping program. The software provides good clear directions with the map as an optional extra rather than the map being the primary guidance of the navigation software.

Microsoft Streets and Trips

Apart from the fact that it has small text and is not very touchscreen friendly, Microsoft's Streets and Trips (see Figure 15-6) is a good piece of mapping software. It shows a trail of breadcrumbs where you have been, which is helpful for putting the next leg of the journey into context. The database for points of interest is unparalleled, and the software will update construction zones via an Internet link. A nice feature for long journeys is that the software will plan into the journey stops for gas based on figures you enter for your vehicle's miles per gallon. One of the disadvantages with using MS Streets and Trips as a navigation program is that it wasn't originally designed as a navigation program; it was designed as a piece of software to allow you to plan a map, print it out, and go. Features have been incorporated to allow you to track your position, but there is no turn-by-turn guidance or and no correction if you deviate from the route.

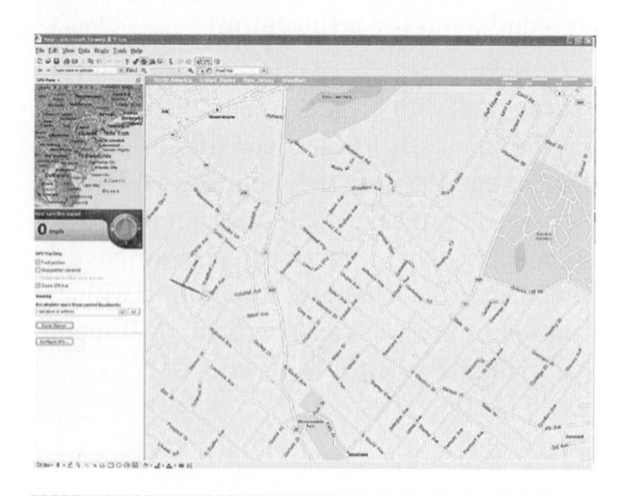

Figure 15-6 *Microsoft Streets and Trips (Courtesy Microsoft)*

Autoroute

Autoroute 2005 is now shipping (see Figure 15-7)! It is based on the software that drives (no pun) Microsoft Map Point. One of the nice features of the new software is the snap routing, which allows you to drag and drop roads to reroute your journey. The software synchronizes well with Microsoft Outlook, which enables you to find a route to any of your saved addresses quickly and easily. You can add vital information to your map by adding pushpins with the drawing tools feature. The software also allows you to plan the amount of

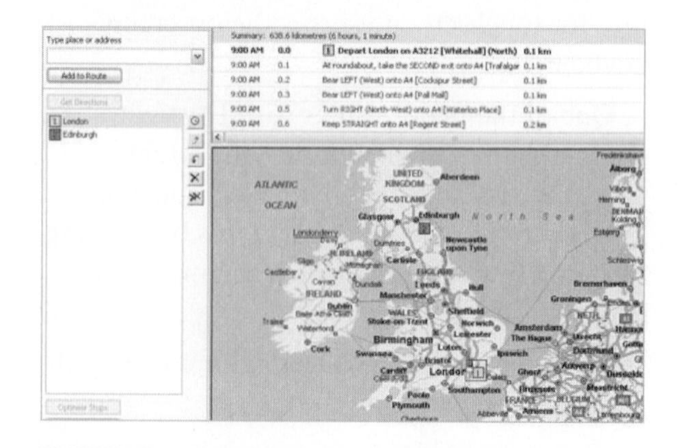

Figure 15-7 *AutoRoute screenshot (Courtesy Microsoft)*

time the journey is going to take with reasonable accuracy.

If you are out on a summer evening and you simply want to drive, but you don't know where, or if you want to take the kids away for a day out, but don't want to drive too far, you can use the drive-time feature. It allows you to plot how far you can drive for any given amount of time.

Microsoft MapPoint

MapPoint is a fully programmable mapping application that can be used to build custom software. It offers fewer features than Streets and Trips, but this is largely because the software is aimed at commercial applications where a customized mapping solution is desired. The software can act as the driver for a navigation program; however, it cannot really compete with some of the other navigation software on the market.

Other Programs

Of particular interest to users in certain countries is a program called GATSO Hunter. This takes GPS data and compares it to a list of known sites of speed cameras (sometimes referred to as revenue generator cameras), which are intended to improve vehicle safety.

In many countries, radar detectors and other methods of safety-camera detectors are being banned. This is bad news for the majority of safe drivers. When tracking a camera with GPS though, you are effectively just comparing your position with a database of locations; therefore, you are doing nothing wrong. A screenshot of the software can be seen in Figure 15-8. The GATSO Hunter software tells you if you are safe and by what distance.

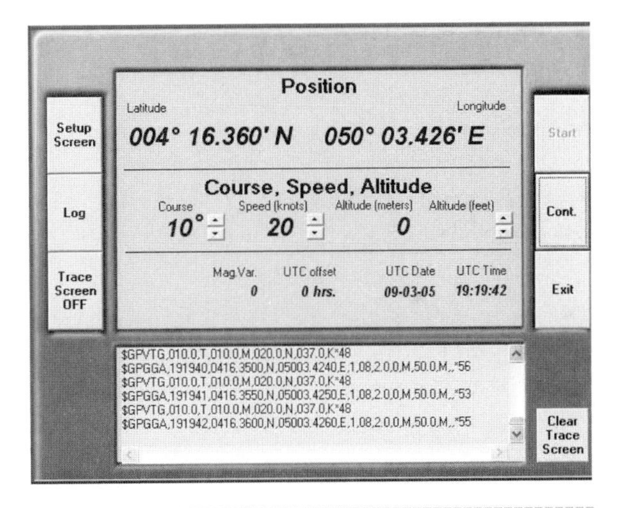

Figure 15-8 *Gatso Hunter screenshot (Courtesy mp3car.com)*

The possibilities are endless . . .

One final thought—a morsel for you to savor and prove *that anything is possible*.

Vespa-Puter!!!

Scared that you haven't got enough room in your car? All it takes is a little ingenuity! Luca Barzelogna, the president of the Italian Vespa-online Association, managed to squeeze a VIA MII motherboard, an 8-inch touchscreen, and a webcam into a VESPA! Wait . . . that's not all! He has also integrated dual hard drives on the step board. This *amazing* vehicle has full GPS functionality, allowing Luca to navigate the Italian streets! Wireless network lets this supercool Vespa browse the Web at wireless hotspots.

And if you thought that was cool, Luca has also enabled this mean machine to use a satellite phone. This has been achieved through ingenuity and the low space requirements of the VIA MII platform.

Go forth and create! Make your vehicle your own!

Appendix A

Suppliers and Helpful Contact Information

Autocom
www.autocomonline.com
info@autocomonline.com
Suppliers of ready-built car PCs

B&B Electronics Manufacturing Company
707 Dayton Road
Ottawa, IL 61350
(815) 433-5100
Suppliers of OBD-II interfaces and software

Car Code
931 Summer Leaf Drive
Saint Peters, MO 63376
Suppliers of OBD-II interfaces and software

Crystal Fontz
www.crystalfontz.com
12412 East Saltese Avenue
Spokane Valley, WA 99216-0357
(509) 892-1200
(888) 206-9720
(509) 892-1203 fax
Suppliers of LCDs

Digital WW
www.digitalww.com
(847) 546-5822
Suppliers of Car PC components

Evation
www.evation.com
irman@evation.com
Suppliers of a range of infrared decoder hardware and software

Formfactors
www.formfactors.org
Design information on ATX power supplies

Hella
Hella KGaA Heuck and Company
Rixbecker Str. 75
59522 Lippstadt
Germany
49 (0) 29 4138 0
Manufacturers of Hella Raintronic Modules

ImagesCo
Images SI Inc.
109 Woods of Arden Road
Staten Island, NY 10312
(718) 966-3594
www.imagesco.com
Suppliers of voice recognition chips

JC Whitney

1 JC Whitney Way

LaSalle, IL 61301

(800) 603-4383

Suppliers of cruise control kits and automobile parts and accessories

Maplin Electronics

National Distribution Centre

Valley Road

Wombwell, Barnsley

South Yorkshire S73 0BS

United Kingdom

Regional stores throughout the UK

Suppliers of electronic components

Matrix Orbital

www.matrixorbital.com/

Suppliers of LCDs

mp3car

www.mp3car.com

(443) 321-4730

store@mp3car.com

Suppliers of car PC components and hardware

National Semiconductor

www.national.com

Suppliers of manufacturer's data sheets

NetOp

www.netop.com

Suppliers of PDA remote control software

OPUS Solutions, Inc.

26941 Cabot Rd, Suite #120

Laguna Hills, CA 92653

(949) 305-4300

(949) 305-4200 fax

info@opussolutions.com

sales@opussolutions.com

tsupport@opussolutions.com

www.opussolutions.com

Suppliers of DC-DC power supplies

Phone Front

www.phonefront.com

Suppliers of phone-control software

Point Research

17150 Newhope Street, Suite 709

Fountain Valley, California 92708-4255

(714) 557-6180 phone

(714) 557-5175 fax

www.pointresearch.com

Suppliers of advanced GPS solutions

RadioShack, USA

(800) 843-7422

www.radioshack.com

Retail stores throughout the United States *and* Canada

Suppliers of all types of electronics parts

Rapid Electronics
Severalls Lane
Colchester, Essex
CO4 5JS
United Kingdom
44 (0) 1206 751166
Suppliers of electronic components, PCB etchant and tin-plating crystals

Rostra Precision Controls
2519 Dana Drive
Laurinburg, NC 28352
(910) 276-4853
(800) 782-3379
www.rostra.com
Suppliers of in-car cruise controls, comfort and security features

RS Components Ltd.
Birchington Road
Corby NN17 9RS
United Kingdom
Suppliers of electronic components

Susquehanna Motorsports
Fleetwood, PA 19522
(610) 944-3233
www.rallylights.com
Suppliers of Hella Raintronic Modules

Trimble
www.trimble.com
Headquarters Phone Number: 11 408 481 8000
(800) 874-6253
Suppliers of Advanced GPS

UBlox
u-blox AG
Zürcherstrasse 68
8800 Thalwil
Schweiz
141 44 722 74 44 phone
141 44 722 74 47 fax
www.u-blox.com
Developers and suppliers of advanced GPS solutions

Wireless Garden, Inc.
339 N. Highway 101, Suite 200
Solana Beach, CA 92075
(888) 509-9434
Suppliers of Cantennas

Appendix B

Glossary

aftermarket any product fitted to a car after it has been purchased from the manufacturer, often manufactured by a company other than the car manufacturer.

ASIC application-specific integrated circuit; an integrated circuit designed to do a particular job.

ATA AT attachment (the original IBM PCs being AT and the hard disk being an attachment); an interface standard for hard disks and computers; more recently we have seen the emergence of SATA.

crosstalk interference caused by electrical conductors being too close in proximity to each other without electromagnetic shielding.

DGPS differential GPS; an improved form of GPS; the errors created by radio signals delayed in their passage through the atmosphere are corrected by a stationary base-station, which broadcasts a signal corresponding to the difference between the signal received and the actual signal.

ECM engine control module; the embedded computer that controls your car's engine; functions may include monitoring engine sensors for normal condition, monitoring fuelling, and controlling emissions.

ECU electronic control unit; refers to embedded computers in cars that the user does not see but perform essential control functions for safety, braking, active suspension control, engine management, and so on.

EGO sensor exhaust gas oxygen sensor; a sensor inserted in the exhaust manifold for monitoring oxygen levels; from this sensor we can derive information about how the engine is running with regards to its fuelling.

GPRS general packet radio service; a system for transmitting data services to mobile phones and devices.

GSM global system for mobile communications; the system used by mobile phones to communicate.

HEGO heated exhaust gas oxygen; variation on EGO where sensor is heated.

IrDa shortened version of infrared data; the standard used by computers and mobile devices for communicating low volumes of data using infrared light.

ISO-EGO isolated EGO sensor (see *EGO sensor*).

Mini-ITX a small motherboard form factor with dimensions of 17cm × 17cm and with low-power consumption requirements.

NMEA National Marine Electronics Association; the name of a GPS protocol taken from the organization that devised it; the NMEA 0138 protocol was originally devised to allow communication between marine electronics and GPS receivers; however, it is now commonly used in many GPS mice.

OBD on-board diagnostics; a system for diagnosing faults with a car's systems using a network of sensors within the car and an ECU; OBD-II is the standard for data exchange within the vehicle.

OBD-II on-board diagnostics, second generation.

OEM original equipment manufacturer.

PATA parallel ATA.

POST power-on self test; the system whereby the motherboard checks itself upon booting for any error; beeps are reported to the user if there is a problem; the pattern of these beeps can be used to diagnose the problem.

SATA serial AT attachment; a high-speed standard for connecting hard disks to computers; it will grow in importance over the next couple of years as PATA drives and controllers are phased out.

SIRF chipset commonly used for GPS mice.

VoIP voice over internet protocol; the technology used to transmit voice calls over the internet by routing the audio information as packets of data over the network.

WiFi wireless fidelity; used to describe high-speed PC networking without wires.

50 Awesome Auto Projects for the Evil Genius

A

B

C

D

E

F

G

H

I

J

L

U

V

W

Z